The Nature and Technology
of Acoustic Space

The Nature and Technology of Acoustic Space

M. Tohyama

NTT Human Interface Laboratories, Musashino-shi, Tokyo, 180 Japan

Present address: Kogakuin University, Tokyo, 163 Japan

H. Suzuki

Ono-Sokki Co., Ltd., Yokohama, 226 Japan

Y. Ando

Kobe University, Kobe, 657 Japan

ACADEMIC PRESS

Harcourt Brace & Company, Publishers

London   San Diego   New York   Boston
Sydney   Tokyo   Toronto

ACADEMIC PRESS LIMITED
24–28 Oval Road
LONDON NW1 7DX

U.S. Edition Published by
ACADEMIC PRESS INC.
San Diego, CA 92101

This book is printed on acid free paper

A catalogue record for this book is available from the British Library

ISBN 0-12-692590-9

Typeset by Doyle Graphics Ltd. Tullamore, Ireland
Printed in Great Britain by Hartnolls Limited, Bodmin, Cornwall

Contents

Preface

Acoustics has been developed with the aid of computer engineering and human science. The science of acoustic space is deeply related to human behaviour. The purpose of this book is to introduce to readers sound field theories and technologies. Signal processing, sound field visualization, sound field control and design methods based on subjective attributes and physiological backgrounds are described in detail. This book is intended for university-level textbook use and also for seminar courses at the graduate student level. It will also be of use for newcomers to be introduced to various fields of acoustic research. Readers are assumed to be familiar to some extent with the fundamentals of the classical theory of room acoustics. Those who are not familiar with these basic theories are recommended to use other appropriate standard texts in conjunction with this text. The authors are hopeful that some of the exercises prepared in this text will be a next step for readers. Since most of the topics described are still under development, the authors welcome any discussion and comment.

Acknowledgements

We are grateful to Professor Dr Manfred R. Schröder, Drittes Physikalisches Institut, Universität Göttingen (Germany), Professor Dr Richard H. Lyon, Mechanical Engineering Department, Massachusetts Institute of Technology (USA), Professor Dr Ken'iti Kido, Department of Computer Science, Chiba Institute of Technology (Japan), Professor Dr Jens Blauert, Lehrstuhl für Allgemeine Elektrotechnik und Akustik, Ruhr-Universität (Germany), and Professor Dr Tammo Houtgast, Institute of Human Factors, TNO and Amsterdam Free University (The Netherlands) for their many scientific discussions and suggestions. Several fundamental studies described in Chapter 5 were performed in cooperation with the Drittes Physikalisches Institut, Universitat Gottingen under the leadership of Professor Schröder. The statistical properties of transfer functions in Chapters 3, 6 and 7 were investigated in detail following Professor Lyon's excellent guidance and very fruitful cooperation. The treatments of speech intelligibility and related topics owe greatly to deep discussions and successful cooperation with Dr Houtgast.

We would thank NTT (Nippon Telegraph and Telephone Corporation) Human Interface Laboratories, Ono Sokki Co. Ltd, and the Acoustic Laboratories of the Graduate School and Faculty of Engineering of Kobe University for allowing us to publish our work. Most of the experimental work was accomplished using measuring equipment, computers and the acoustical facilities of NTT, Ono-Sokki, and Kobe University.

We also acknowledge all of our colleagues who encouraged and helped us to accomplish our experiments, calculations and final documentation. We thank Dr Takahiko Ono (President of Ono-Sokki Co. Ltd) and Dr S. Furui (NTT Human Interface Laboratories) for their able guidance and daily encouragement. We thank Mr Tsunehiko Koike (NTT Advanced Technology Corporation) for his excellent cooperation. Finally, we are most grateful to Dr P. A. Nelson (Southampton University, UK) for his suggestion of publishing through Academic Press.

1

Introduction

Acoustic theory and technology have been developed on the basis of computer engineering and human science. This book describes control and design methods for desirable sound fields according to modern acoustic theory and technology. Initially, the theoretical aspects of wave propagation are mathematically treated. Visualization of sound fields using sound energy flow helps us to understand the physical properties of the sound fields. Then, the perplexing problems of subjective attributes of sound fields as well as human brain activity are discussed. After the determination of preferred conditions of sound fields, control technologies are demonstrated for sound field optimization.

In Chapter 2, various signal processing methods that will be used in the later chapters are described. As this book is not written specifically for signal processing applications, descriptions are limited to what is necessary and important for understanding in the following chapters. The first two sections are devoted to Fourier analysis, the main method of present-day signal analysis. Since the continuous Fourier transform is covered in undergraduate courses, only basic properties and transform pairs of the continuous Fourier transform are listed in Section 2.1. Readers not familiar with the continuous Fourier transform should refer, for example, to Papoulis (1962) and Weaver (1989).

In Section 2.2, the discrete Fourier transform is described. Since the advent of the fast Fourier transform together with the recent advancement of the digital computer, the Fourier transform has gained engineering value in signal analysis. However, it is important to understand the differences as well as the commonalities between the continuous and discrete Fourier transforms. When a signal is captured digitally with a finite duration, a time window (rectangular in this case) is automatically introduced. In this section, significant effort will be centred on explaining the effects of windowing in the discrete Fourier transform.

Section 2.3 introduces the Hilbert transform, which plays an important role in mathematically describing the properties of signals such as causality and minimum phase shift. The analytic signal, which has only positive frequency components, is also explained by use of the Hilbert transform.

In the final section of Chapter 2, several methods of signal analysis on the time and frequency $(t-f)$ plane are described. The Fourier transform and the inverse Fourier transform are methods of looking at a signal in either the time or frequency domain. In some cases, however, we may wish to quantitatively analyse a frequency spectrum at a specific time. Because of the Uncertainty Principle, there is no perfect method of

1

analysing the signal on the $t-f$ plane with unlimited resolution. The Wigner distribution and similar variations are used for this purpose. The Wigner distribution and its application examples are introduced here.

Chapter 3 describes sound propagation in a room. Readers are assumed to be familiar with the fundamentals of the wave theory for rooms. Readers not familiar with room acoustic theory are recommended to use standard textbooks, for example, Kuttruff (1991) for reference while reading this text.

The first two sections of Chapter 3 cover typical examples of reflection at a boundary. Section 3.1 introduces interference patterns induced by reflection at a wall. The frequency characteristic of a signal path including a single reflection from a boundary is a typical example of digital filter design. We introduce here the concept of delay and the periodicity of the frequency characteristics. Section 3.2 discusses the effects of boundaries on the sound power output of a sound source. The transmission loss of a single wall is described as an example that demonstrates the changes of radiation impedance dependent on the frequency of the wall vibration. Section 3.3 reviews room reverberation theory. The reverberation theory of nonexponential decay fields is investigated as well as conventional theories. A measurement example confirming the reverberation theory is illustrated.

Sections 3.4, 3.5 and 3.6 are devoted to the statistical analysis of random sound fields. In Section 3.4, the fundamentals of the statistics of random sound fields are introduced. Section 3.5 mainly describes the distribution statistics of the poles and zeros of transfer functions in reverberant sound fields. Although poles are located on the pole-line parallel with the real frequency line, the zeros are distributed according to a Cauchy distribution in the complex frequency plane. Study of the locations of the zeros of the transfer functions may yield various application areas of acoustics technology.

Section 3.6 introduces the theory of SPA (statistical phase analysis) of transfer functions. SEA (statistical energy analysis) is a well-known and powerful design tool. However, the use of advanced technology (as, for example, in machine diagnostics) allows one to control the sound waveform, which is sensitive to the phase property of the sound fields. The theory of SPA based on the distribution statistics of poles and zeros may be useful in estimating the phase behaviour of random sound fields. The final section, 3.7, describes the MTF (modulation transfer function) in a reverberant space. The MTF is derived from the impulse response of a reverberant space, but it can be applied to estimate speech intelligibility under reverberant conditions.

Following theoretical descriptions of wave propagation in rooms and presentation of experimental results, Chapter 4 will concentrate on visualization of the sound field. One of the conventional methods of sound field visualization is the pressure contour display. This provides various types of sound field information such as source location, location of reflections, and existence of standing waves. However, as will be seen in Chapter 4, the pressure distribution does not always give sufficient information for complete understanding of the sound field. Another common method is ray tracing. This method is intuitive and useful for the design of the room since it gives an estimation of the sound field when the source and room conditions are known. However, it does not give details of the sound field, owing to its inherent simplification of the processes of the wave propagation and the reflections at boundaries.

The main method, used in Chapter 4, for visualization of the sound field is the intensity method, which was mostly developed in the last decade. As a sound wave propagates, energy flows through an acoustic medium. Sound radiation mechanisms from loudspeakers or any other type of sound source are better understood by visualizing the sound field in a room or around sources. The advantage of the intensity method is that it can be used in both theoretical and experimental evaluations of the sound field.

In Section 4.1, definitions of the instantaneous and time-average sound intensities are given. The basics of the most commonly used two-microphone method are explained. The two microphone outputs are used to obtain the sound intensity in two ways, the direct integration method and the cross spectrum method.

In Section 4.2, visualizations of various types of sound fields such as propagating waves, standing waves and wave fields in a reverberation room are given. The existence of the intensity vortex is shown theoretically and experimentally. Measurement results around a polyurethane foam absorber placed on a rigid floor of a hemi-anechoic room are also given. Sound fields around sources such as monopoles, flat, convex and concave radiators, loudspeakers and rectangular plates are dealt with in Section 4.3. Interactions among coherent monopoles are visualized by the intensity method. The intensity fields around convex and concave domes clearly explain the differences in radiation characteristics from them.

Intensities other than the commonly used time-average sound intensity are shown in Section 4.4. They are envelope intensity and the instantaneous sound intensity spectrum. The latter is the application of the cross Wigner distribution to the sound intensity. Vibration intensity is an energy flow per unit width in a structure. This method can be used to visualize energy flow in beams and plates.

In Chapter 5, we discuss subjective attributes of sound fields. Preferred conditions of sound fields can be described by the four orthogonal factors including temporal and spatial criteria. In the first section of Chapter 5, the theory of subjective preference with a limited number of orthogonal acoustic factors of a sound field in a room is reviewed. In order to confirm a model of auditory brain functioning, slow vertex responses (SVR) are recorded and analysed in Section 5.2. It is remarkable that, according to the temporal and spatial factors, hemispherical dominance appears in the amplitude of SVR, and latencies of SVR, especially N_2 latency, are discovered to be related to the subjective preference judgement.

From records of the auditory brainstem responses (ABR) as well as SVR, a possible mechanism of the spatial factor IACC (interaural cross correlation function) is discussed. In Section 5.3, ABR as a function of the horizontal angle of a sound source's location relative to that of a listener is demonstrated, and the ABR amplitude relating to the IACC is discussed.

In accordance with above-mentioned subjective and electrophysiological responses, Section 5.4 proposes a model of the auditory–brain system with hemispheric specialization for temporal and spatial acoustic factors. Some important subjective attributes are reviewed in relation to the autocorrelation function (ACF) of source signals and the magnitude of the interaural cross correlation function (IACC).

In Section 5.5, in terms of such a theory with a limited number of orthogonal factors, the individual differences in subjective preference are described and are discussed in respect to individual preferences for certain seat locations in a concert hall. In the final section of Chapter 5, several other important subjective attributes are approximately expressed by the envelope function of the ACF and the IACC. The subjective attributes of loudness, coloration, echo disturbance and preferred delay for a performer are related to the envelope of the ACF. Subjective diffuseness and even speech intelligibility and clarity are desribed as a function of IACC, in addition to temporal criteria of the sound field.

Chapter 6 begins with the passive control of sound power transmission. Section 6.1. describes loudspeaker location in a vehicle as an example of passive control of the power transmission. Section 6.1.2 briefly reviews the methods of power measurement of a sound source. Section 6.2 outlines the technologies developed for sound field reproduction using two-channel or multichannel loudspeakers. Expansion of the listening area for good sound image localization, the relationship between subjective diffuseness and interaural cross correlation, and sound field simulation using multi-channel loudspeakers are described.

Sections 6.3 and 6.4 deal with the applications of signal processing for acoustic technology. Section 6.3 introduces inverse filtering and other dereverberation technologies for source waveform recovery in a reverberant space. A pulse-like source waveform recovery is important for machine diagnostics according to the source signal signature. If the source waveform can be extracted from reverberant responses by dereverberation, the change in the source signature is useful information for the machine diagnostics. Dereverberation of speech signals is applied to improve the quality of speech communication under reverberant conditions.

Section 6.4 shows examples of adaptive and active control for sound fields. An echo-canceller is an indispensable tool for a hands-free telecommunication system. Echo cancelling can be realized using the adaptive technology. A new number-theory approach for active sound field control is illustrated. Sound pressure signals at n points in a space can be controlled if we can use $n + 1$ loudspeakers and obtain the transfer functions of all pairs of loudspeakers and microphones.

In Chapter 7, technologies and examples for sound field control in a concert hall are demonstrated. Section 7.1 describes sound reinforcement for speech intelligibility in a reverberant space. Section 7.2 is related to the temporal aspects of sound localization under reverberant conditions. The interaural cross correlation for nonstationary signals is introduced. A series of subjective experiments on the role of precedence effects on sound image localization are described.

Section 7.3 describes digital technologies employed in concert hall acoustics. Sound transmission measurements and visualization of virtual sources distributions in a concert hall are developed. Examples of equalizing filters are illustrated for a public address system and howling cancellation technologies are mentioned.

For the purpose of optimizing the sound field for given subjective attributes, Section 7.4 describes the design study of concert halls. The method of control based on the temporal criteria and IACC is illustrated by considering the design of a special reflector and of an electroacoustic system. A seat selection system is proposed that is based on a listener's preference.

2

Signal Analysis

What we hear in the sound field is a fluctuation of the air pressure, which we call a 'signal' when it is captured as an electrical output by a microphone. Therefore, if the microphone is as good as the human ear, electrical signals should contain full information with respect to differences among perceived sounds. If the signal is analysed in the same way as in the human brain, we can construct an ideal sound analyser to evaluate the subjective quality of the sound. We are still far from that goal, but recent developments in signal analysis have brought us one step closer.

In this chapter, various signal processing methods are described that will be used in the following chapters. Since this volume is not a signal processing book, descriptions are limited mainly to those that are necessary in understanding the remainder of the book. In the first section, the continuous Fourier transform is briefly reviewed. It is assumed, however, that the reader is familiar with this method and, therefore, some properties of the Fourier transform and common transform pairs are listed for convenience without proof. The second section introduces the discrete Fourier transform, which is necessary for application of concepts of the continuous Fourier transform to digital signal processing. Emphasis is placed on explaining the basics of the discrete Fourier transform and especially the effects of time windows. The third section describes the importance of the Hilbert transform in understanding basic properties of signals and systems such as causality, minimum phase shift and analyticity. The last section is devoted to introducing several signal analysis methods in the time–frequency domain. Understanding the concept of the Wigner distribution on the time–frequency domain is the main concern of this section.

2.1. Continuous Fourier Transform

Since the publication of *The Analytical Theory of Heat* in 1822 by Joseph Fourier, Fourier analysis has played a significant role in mathematical physics. This method became the most powerful technique in the field of signal analysis after the fast Fourier transform algorithm (FFT) was conceived by Cooley and Tukey (1965). In this section, only the definitions, the properties of the continuous Fourier transform, and the transform pairs of some functions are presented. Readers who are unfamiliar with the theory should refer to Papoulis (1962) and Weaver (1989).

2.1.1. Definition and Properties of the Fourier Transform

The forward and inverse Fourier transform pairs are defined, respectively, as

$$X(f) = \int_{-\infty}^{\infty} x(t)e^{-j2\pi ft}\, dt \tag{2.1.1}$$

$$x(t) = \int_{-\infty}^{\infty} X(f)e^{j2\pi ft}\, df \tag{2.1.2}$$

A sufficient condition for the existence of the above integrals is absolute integrability, that is,

$$\int_{-\infty}^{\infty} |x(t)|\, dt < \infty \tag{2.1.3}$$

and

$$\int_{-\infty}^{\infty} |X(f)|\, df < \infty \tag{2.1.4}$$

Absolutely integrable transform pairs satisfy the properties shown in Table 2.1.1.
Furthermore, if $x(t)$ is absolutely integrable, $X(f)$ has the following properties:

(a) If $x(t)$ is real, then $X(-f) = X^*(f)$.
(b) If $x(t)$ is real and even, then $X(f)$ is real and even.
(c) If $x(t)$ is real and odd, then $X(f)$ is imaginary and odd.
(d) If $x(t)$ is imaginary and even, then $X(f)$ is imaginary and even.
(e) If $x(t)$ is imaginary and odd, then $X(f)$ is real and odd.

Table 2.1.1. Properties of the Fourier transform

Property	$x(t)$	$X(f)$
Linearity	$ax_1(t) + bx_2(t)$	$aX_1(f) + bX_2(f)$
Time translation (first shifting theorem)	$x(t - a)$	$X(f)e^{-j2\pi af}$
Frequency translation (second shifting theorem)	$x(t)e^{j2\pi f_o t}$	$X(f - f_0)$
Modulation	$x(t)\cos 2\pi f_c t$	$\frac{1}{2}[X(f - f_c) + X(f + f_c)]$
Scaling	$x(at)$	$\dfrac{X(f/a)}{|a|}$
Transform of the derivative	$\dfrac{d^n}{dt^n}[x(t)]$	$[j2\pi f]^n X(f)$
	$[-j2\pi t]^n x(t)$	$\dfrac{d^n}{df^n}[X(f)]$
Transform of the integration	$\displaystyle\int_{-\infty}^{t} x(t')\, dt'$	$(j2\pi f)^{-1} X(f)$
Transform of a convolution	$\displaystyle\int_{-\infty}^{\infty} x_1(\tau)x_2(t - \tau)\, d\tau$	$X_1(f)X_2(f)$
Transform of a correlation	$\displaystyle\int_{-\infty}^{\infty} x_1^*(\tau)x_2(t + \tau)\, d\tau$	$X_1^*(f)X_2(f)$

2.1.2. Fourier Transform as a Distribution

Periodic signals such as $\cos 2\pi ft$ are not absolutely integrable, and therefore those functions are not found in tables of Fourier transforms pairs in most of books. However, the Fourier transform of $\cos 2\pi ft$ can be obtained if we introduce the concept of distribution theory and the impulse function (Papoulis, 1962; Weaver, 1989). The impulse function is defined as

$$\int_{-\infty}^{\infty} \delta(t - t_0)x(t)\, \mathrm{d}t = x(t_0) \tag{2.1.5}$$

where the integral and $\delta(t - t_0)$ is considered as a process for assigning $x(t_0)$ to $x(t)$. Then

$$\int_{-\infty}^{\infty} \delta(f)e^{j2\pi ft}\, \mathrm{d}f = 1 \tag{2.1.6}$$

Therefore we obtain the Fourier transform pair:

$$1 \leftrightarrow \delta(f) \tag{2.1.7}$$

In the same way,

$$e^{j2\pi f_0 t} \leftrightarrow \delta(f - f_0) \tag{2.1.8}$$

Since

$$\cos 2\pi f_0 t = \tfrac{1}{2}(e^{j2\pi f_0 t} + e^{-j2\pi f_0 t}) \tag{2.1.9}$$

we obtain the Fourier transform pair:

$$\cos 2\pi f_0 t \leftrightarrow \tfrac{1}{2}[\delta(f - f_0) + \delta(f + f_0)] \tag{2.1.10}$$

Similarly,

$$\sin 2\pi f_0 t \leftrightarrow \frac{1}{2j}[\delta(f - f_0) - \delta(f + f_0)] \tag{2.1.11}$$

2.1.3. Examples of Transform Pairs

Examples of Fourier transform pairs of basic functions are shown here.

(1) *Rectangular pulse* (width $2T$) (Fig. 2.1.1):

$$x(t) = p_T(t) = \begin{cases} 1 & |t| < T \\ 0 & |t| > T \end{cases} \qquad X(f) = \frac{\sin 2\pi Tf}{\pi f} \tag{2.1.12}$$

Similarly,

$$x(t) = \frac{\sin 2\pi Ft}{\pi t} \qquad X(f) = p_F(f) = \begin{cases} 1 & |f| < F \\ 0 & |f| > F \end{cases} \tag{2.1.13}$$

(2) *Pulse-modulated cosine* (Fig. 2.1.2):

$$x(t) = p_T(t)\cos 2\pi f_0 t \qquad X(f) = \frac{\sin 2\pi(f - f_0)T}{2\pi(f - f_0)} + \frac{\sin 2\pi(f + f_0)T}{2\pi(f + f_0)} \tag{2.1.14}$$

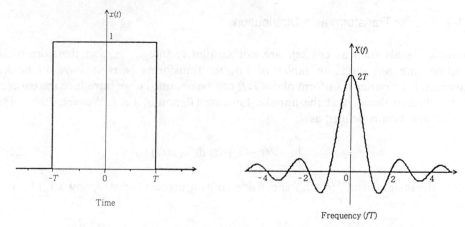

Fig. 2.1.1. Fourier transform pair of a rectangular pulse (width 2T).

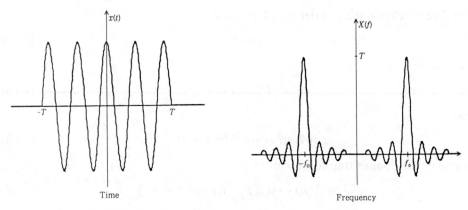

Fig. 2.1.2. Fourier transform pair of a pulse-modulated cosine.

(3) *Two rectangular pulses* (distance 4T) (Fig. 2.1.3):

$$x(t) = p_T(t + 2T) + p_T(t - 2T) \qquad X(f) = \frac{\sin 2\pi Tf}{\pi f} (e^{-j2T2\pi f} + e^{+j2T2\pi f})$$

$$= \frac{2 \sin 2\pi f T}{\pi f} \cos 4\pi Tf \tag{2.1.15}$$

(4) *Triangular pulse* (Fig. 2.1.4):

$$x(t) = \begin{cases} 1 - |t|/T & |t| < T \\ 0 & |t| > T \end{cases} \qquad X(f) = \frac{\sin^2(\pi Tf)}{\pi^2 Tf^2} \tag{2.1.16}$$

(5) *Exponential function* (Fig. 2.1.5):

$$x(t) = e^{-\alpha t} U(t) \ (U(t): \text{ unit step function}) \qquad X(f) = \frac{1}{\alpha + j2\pi f}$$

$$\tag{2.1.17a}$$

$$= \frac{1}{\sqrt{\alpha^2 + (2\pi f)^2}} e^{-j \tan^{-1}(2\pi f/\alpha)}$$

$$\phi(f) = -\tan^{-1}(2\pi f/\alpha) \tag{2.1.17b}$$

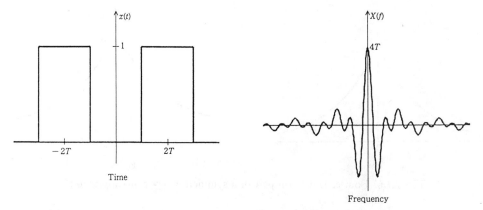

Fig. 2.1.3. Fourier transform pair of two rectangular pulses.

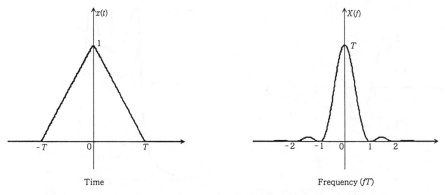

Fig. 2.1.4. Fourier transform pair of a triangular pulse (width $2T$).

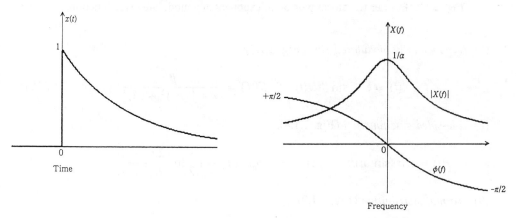

Fig. 2.1.5. Fourier transform pair of an exponential function.

(6) *Symmetric exponential function* (Fig. 2.1.6):

$$x(t) = e^{-\alpha|t|} \qquad X(f) = \frac{2\alpha}{\alpha^2 + (2\pi f)^2} \qquad (2.1.18)$$

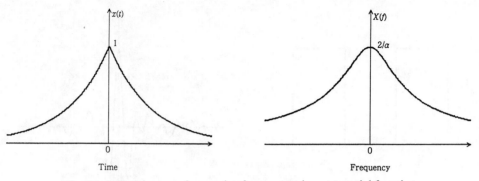

Fig. 2.1.6. Fourier transform pair of a symmetric exponential function.

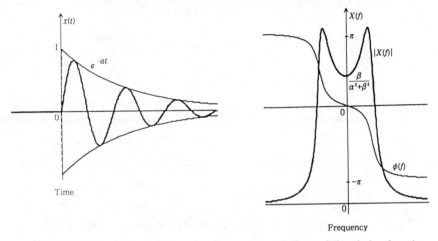

Fig. 2.1.7. Fourier transform pair of an exponentially modulated sine function.

(7) *Exponentially modulated sine* (Fig. 2.1.7):

$$x(t) = e^{-at} \sin \beta t U(t) \qquad X(f) = \frac{\beta}{(\alpha + j2\pi f)^2 + (\beta)^2} \qquad (2.1.19)$$

(8) *One-sided sine function* (Fig. 2.1.8):

$$x(t) = U(t)(\sin at)/t \qquad X(f) = \frac{\pi}{2} P_{a/2\pi}(f) + j\frac{1}{2} \ln \left| \frac{2\pi f - a}{2\pi f + a} \right| \qquad (2.1.20)$$

(9) *Normal distribution* (Fig. 2.1.9):

$$x(t) = e^{-\alpha t^2} \qquad X(f) = \sqrt{\frac{\pi}{\alpha}} e^{-\pi^2 f^2/\alpha} \qquad (2.1.21)$$

(10) *Delta function* (Fig. 2.1.10):

$$x(t) = \delta(t) \qquad X(f) = 1 \qquad (2.1.22)$$

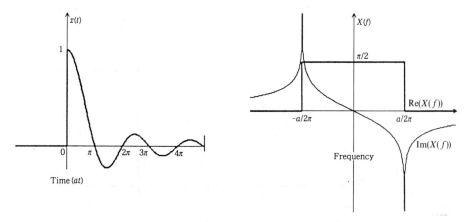

Fig. 2.1.8. Fourier transform pair of a one-sided sinc function.

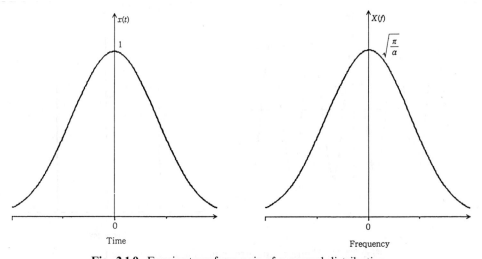

Fig. 2.1.9. Fourier transform pair of a normal distribution.

(11) *Cosine* (Fig. 2.1.11):

$$x(t) = \cos 2\pi f_0 t \qquad X(f) = \tfrac{1}{2}[\delta(f - f_0) + \delta(f + f_0)] \tag{2.1.23}$$

(12) *Sine* (Fig. 2.1.12):

$$x(t) = \sin 2\pi f_0 t \qquad X(f) = \frac{1}{2j}[\delta(f - f_0) - \delta(f + f_0)] \tag{2.1.24}$$

(13) *Sign (signum) function* (Fig. 2.1.13):

$$x(t) = \begin{cases} -1 & t < 0 \\ +1 & t > 0 \end{cases} \qquad X(f) = \frac{1}{j\pi f} \tag{2.1.25}$$

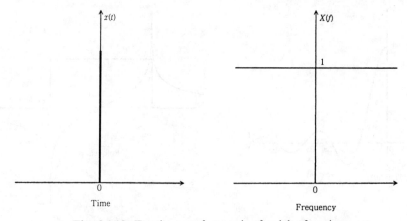

Fig. 2.1.10. Fourier transform pair of a delta function.

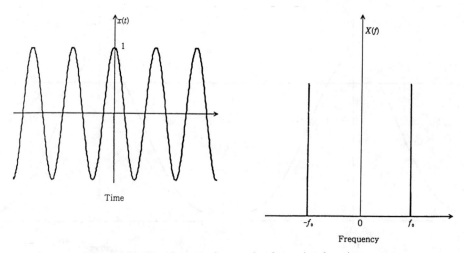

Fig. 2.1.11. Fourier transform pair of a cosine function.

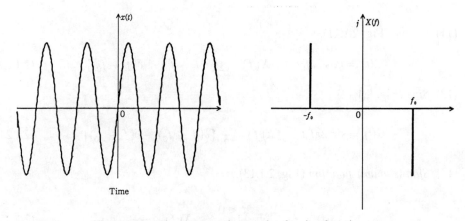

Fig. 2.1.12.   Fourier transform pair of a sine function.

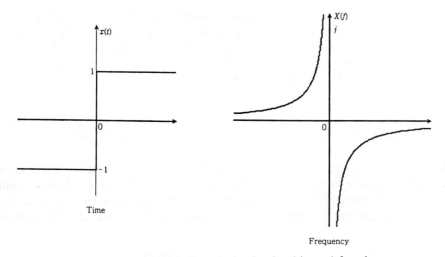

Fig. 2.1.13. Fourier transform pair of a sign (signum) function.

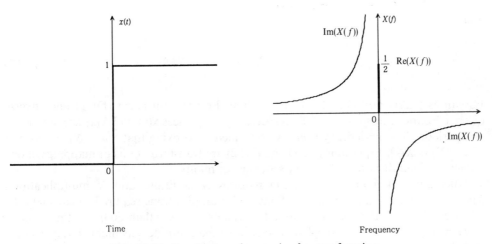

Fig. 2.1.14. Fourier transform pair of a step function.

(14) *Step function* (Fig. 2.1.14):

$$x(t) = U(t) = \begin{cases} 0 & t < 0 \\ 1 & t > 0 \end{cases} \qquad X(f) = \frac{1}{2}\delta(f) + \frac{1}{j2\pi f} \qquad (2.1.26)$$

Note that if $X(\omega)$ is given instead of $X(f)$, $X(f)$ is given simply by replacing ω by $2\pi f$. If $X(f)$ is given as a Fourier transform of $x(t)$, a transform of $x(2\pi f_0 t)$ is given by $X(f/2\pi f_0)/2\pi|f_0|$ (scaling).

2.2. Discrete Fourier Transform

The continuous Fourier transform is defined in the infinite time or frequency domain and the integration is possible only for limited functions. For an arbitrary time or frequency function, the integration must be performed numerically, and accordingly the time and frequency spans must be finite. A discrete transform or mapping that has analogies to the continuous Fourier transform is called a discrete Fourier transform (DFT) (Cooley and Tukey, 1965; Gold and Rader, 1969; Bendat and Piersol, 1971; Oppenheim and Shafer, 1975). The DFT method is the most useful and commonly used signal analysis method that is presently available.

2.2.1. DFT and FFT

A signal $x(t)$ and its spectrum $X(f)$ are related to each other by the Fourier transform and inverse Fourier transform (Eqs (2.1.1) and (2.1.2)). The continuous Fourier transform in the infinite time and frequency domains cannot be handled by a digital computer. A fundamental theory of the digital signal analysis is based on the equations

$$X(k) = \sum_{n=0}^{n=N-1} x(n)\, e^{(-j2\pi kn/N)} \tag{2.2.1}$$

and

$$x(n) = \frac{1}{N} \sum_{k=0}^{k=N-1} x(k)\, e^{(+j2\pi kn/N)} \tag{2.2.2}$$

Equations (2.2.1) and (2.2.2) are the discrete Fourier transform (DFT) and inverse discrete Fourier transform (IDFT), respectively. Sequences $x(n)$ and $X(k)$ are both finite and periodic. The periodicity is easily confirmed by proving that $x(n + N) = x(n)$ and $X(k + N) = X(k)$. Equations (2.2.1) and (2.2.2) are transformations or mappings of one sequence of numbers into the other sequence of numbers.

Calculation of $X(k)$ in Eq. (2.2.1) requires N additions and N multiplications. Therefore, the total number of additions and multiplications required to calculate the whole DFT is N^2. If N becomes large (for example, larger than 500), the time needed to calculate the DFT is not negligible even with the computers available today. Cooley and Tukey (1965) invented an algorithm to calculate the DFT with $(N/2)\log_2 N$ additions and multiplications when N is given by $N = 2^m$ where m is a positive integer. If N is equal to 1024 ($m = 10$), for example, the number of additions and multiplications is reduced by a factor of $1/205$. This method is called the fast Fourier transform (FFT). Nowadays, since the FFT is very common, it is sometimes used with the same meaning as the DFT. Computer programs for FFT in different languages are available today that can be used on personal computers as well as large mainframe computers. Also, a specially designed signal processor based on the FFT method is now available. An explanation of the FFT algorithm will be omitted here since it is described in most signal processing books.

Analogies between the continuous and discrete Fourier transforms are shown by simple examples of DFT. Figure 2.2.1 shows transformations of the digitized impulse, cosine, and sine signals for the case of $N = 8$. If $\{x(n)\} = \{1, 0, 0, 0, 0, 0, 0, 0\}$ then $\{X(k)\} = \{1, 1, 1, 1, 1, 1, 1, 1\}$ where $\{\ \}$ indicates a sequence of numbers. This result corresponds to Fig. 2.1.10. For the delayed impulse $\{x(n)\} = \{0, 1, 0, 0, 0, 0, 0, 0\}$, $X(k) = e^{-j2\pi k/8}$, that is, $X(k)$ is phase shifted proportionally to k with a constant

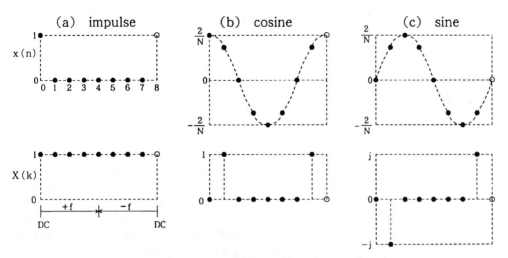

Fig. 2.2.1. Examples of discrete Fourier transforms.

$2\pi/N\,(N=8)$. This is analogous to the time shift property in Table 2.1.1. For the cosine and sine signals, nonzero components appear only for $k=1$ and 7 (or -1). Values of $X(k)$ for $N/2 < k \leqslant N-1$ are considered to represent the frequency spectrum of the time signal $\{x(n)\}$ at negative frequencies. Figure 2.2.2 is a circle showing the periodicity of the DFT for $N=8$. Positions of $X(k+iN)$, $i=\ldots-1,0,1,\ldots$ overlap on the circle. $X(0)$ is considered to represent a d.c. (direct current) component since it is a sum of $x(n)$, $n=0,1,2,\ldots,N-1$. The position of $X(N-k)$ on the circle is symmetric with that of $X(k)$ with respect to the horizontal axis, and its value is a complex conjugate of $X(k)$ if $\{x(n)\}$ is real (see the symmetric property of Section 2.1). The component at $n=N/2$ can be considered either as a positive or a negative frequency component.

In order to give physical meanings to the sequences of numbers $x(n)$ and $X(k)$, we will consider that $x(n)$, $n=0,1,2,\ldots,N-1$, is a time signal sampled at a time interval

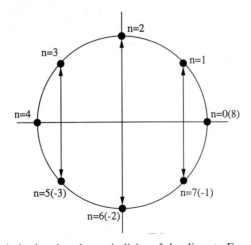

Fig. 2.2.2. A circle showing the periodicity of the discrete Fourier transform.

(period) Δt. Then relations between various parameters are given as follows:

(1) Sampling period Δt (0.390 625 ms)
(2) Number of samples N (1024)
(3) Sampling frequency (rate) $f_s = 1/\Delta t$ (2.56 kHz)
(4) Window length $T = N \Delta t$ (0.4 s)
(5) Frequency resolution $\Delta f = 1/T$ (2.5 Hz)
(6) Displayed maximum frequency $f_{max} = 0.390\,625 \times f_s = 400 \times \Delta f$ (1 kHz)

If we assume two independent values $\Delta t = 0.390\,625$ ms and $N = 1024$, then the values of the related parameters are given as shown in parentheses. In most commercially available FFT analysers, if $N = 1024$, only the first 401 components (so called lines) are displayed. In the above example, the displayed upper frequency (maximum frequency) is 1 kHz. The negative frequency components do not have to be displayed since they are complex conjugates of corresponding positive frequency components as long as $\{x(n)\}$ is real. (These undisplayed components must also be taken care of when the IFFT is executed.) Since $f_s = 2.56$ kHz, the sampled signal should not contain frequency components above 1.28 kHz $(0.5f_s)$ because of Shannon's sampling theorem (Shannon, 1949). In the above example, the displayed upper frequency $(0.390\,625f_s)$ is lower than half of the sampling frequency, satisfying Shannon's sampling theorem. In order to avoid the aliasing problem, aliasing filters with amplitude and phase responses such as shown in Fig. 2.2.3 are used (f_{max} at 1 kHz). The amplitude is reduced by only 40 dB at half of the sampling frequency. This filter gives a reduction of aliased components by approximately 80 dB at f_{max}, which seems to be sufficient for most cases.

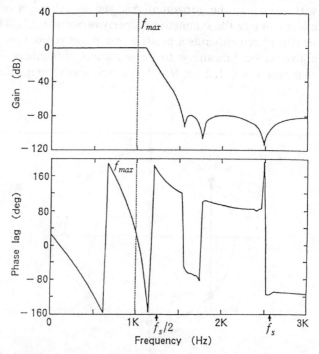

Fig. 2.2.3. Amplitude (top) and phase (bottom) responses of an aliasing filter.

It should be noted, however, that the phase of the frequency spectrum is changed significantly even in the range of analysis. This is not a problem if only the power spectrum is concerned, or a frequency response function is obtained by use of a two-channel FFT analyser with two almost identical aliasing filters.

2.2.2. Time Windows

Since $x(n)$ and $X(k)$ are periodic (with a window length T), there arises a problem that does not exist in the continuous Fourier transform. For example, in Fig. 2.2.1, the cosine and sine signals are sampled so that the start and end of $x(n)$ are connected smoothly. Therefore, the shape of the original continuous signal is preserved and its Fourier transform $X(k)$ has its nonzero components at two (positive and negative) frequencies as the original signal does. However, if T is increased by 50%, for example, the sampled periodic signal looks very different from the original signal and the resultant $X(k)$ has nonzero components at more than two frequencies. This is due to the fact that the present sampling looks at the signal through a rectangular (with no weighting) time window. The effect of the windowing is shown in Fig. 2.2.4. A sinusoidal signal with frequencies at 200 Hz and 206.25 Hz is sampled by use of a rectangular

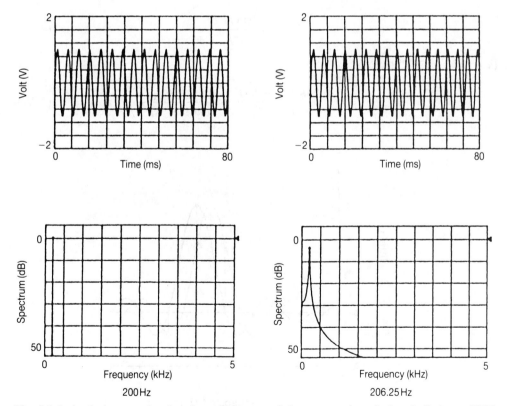

Fig. 2.2.4. Analysis example of a sinusoid by use of the rectangular window built in an FFT analyser.

window with $f_s = 12.8\,\mathrm{kHz}$ and $N = 1024$ ($\Delta f = 12.5\,\mathrm{Hz}$). The top and bottom figures show the sampled waveforms and frequency spectra ($|X(k)|$) for $f = 200\,\mathrm{Hz}$ (left) and $206.25\,\mathrm{Hz}$ (right), respectively. The sampled waveform is continuous if $f = 200\,\mathrm{Hz}$ and a nonzero component of $|X(k)|$ appears only at $200\,\mathrm{Hz}$ and the measured value is very close to the true value ($0\,\mathrm{dB}$). However, for $f = 206.25\,\mathrm{Hz}$, the sampled waveform shows a severe discontinuity and the largest amplitude is $-3.79\,\mathrm{dB}$ at $200\,\mathrm{Hz}$. This simple example indicates the difficulty of FFT analysis and appropriate care must be taken to avoid this kind of problem.

Let us discuss this problem first in the continuous time and frequency domains. The time-windowed signal $x_w(t)$ generated by a window function $w_r(t)$ is given by

$$x_w(t) = x(t)w_r(t) \tag{2.2.3}$$

The frequency spectrum of this signal is

$$X_w(f) = X(f) \otimes W_r(f) \tag{2.2.4}$$

where $\otimes$ denotes a convolution. If $x(t)$ is a cosine with unit amplitude and $w_r(t)$ is a rectangular window function with length T, then

$$X(f) = \tfrac{1}{2}(\delta(f - f_0) + \delta(f + f_0)) \tag{2.2.5}$$

and

$$W_r(f) = \frac{\sin \pi f T}{\pi f} \tag{2.2.6}$$

(see Eq. 2.1.12) Therefore

$$X_w(f) = \frac{1}{2}\left(\frac{\sin \pi(f - f_0)T}{\pi(f - f_0)} + \frac{\sin \pi(f + f_0)T}{\pi(f + f_0)}\right) \tag{2.2.7}$$

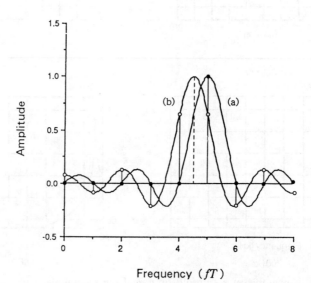

Fig. 2.2.5. Sine functions that illustrate the effect of the rectangular window.

If the signal is discretized with a window length T and a sampling frequency f_s, the discrete frequencies are $0, \pm 1/T, \pm 2/T, \dots, \pm f_s/2$. Furthermore, if $f_0 = m/T$, complete m periods are included in the window and only in this case does the sampled time sequence keep the original smooth sinusoidal shape. In this case, a nonzero component appears only at frequency m/T with zero at all other frequencies since $\sin \pi[(f - f_0)T]/\pi(f - f_0)$ is equal to T for $f = m/T$ and zero for other discrete frequencies. Figure 2.2.5(curve a) illustrates this for the case $m = 5$ in the positive frequency region. If $m = 4.5$, however, there is an infinite number of nonzero components, as can be seen from Fig. 2.2.5(curve b). (In the figure, only the first term of Eq. (2.2.7) is considered.) This problem is caused by the periodicity of the sampled sequence, which sometimes gives rise to a discontinuity between the start and end points of the sequence. The rectangular window should be used for periodic signals if the window length can be made equal to an integer multiple of the signal period. It should also be used for transient signals that are equal to zero at the start and end points. Obviously there is no power reduction due to the rectangular window.

The most commonly used window for solving the discontinuity problem is the Hanning window (raised cosine), which is defined by

$$w_n(t) = \frac{1}{2}\left(1 + \cos 2\pi \frac{t}{T}\right) = \cos^2 \pi \frac{t}{T} \qquad (-T/2 \leqslant t \leqslant T/2)$$

$$= 0 \text{ (otherwise)} \tag{2.2.8}$$

$$W_n(f) = \frac{\sin \pi f T}{2\pi f} + \left(\frac{\sin \pi(f - 1/T)T}{2\pi(f - 1/T)} + \frac{\sin \pi(f + 1/T)T}{2\pi(f + 1/T)}\right) \tag{2.2.9}$$

The Hanning window functions $w_n(t)$ and $W_n(f)$ are shown in Fig. 2.2.6. Since the Hanning window gradually decreases to zero at both ends, the discontinuity between the start and end points is removed. Of course, the shape of the original signal is always distorted. $W_n(f)$ is equal to $0.5T$ and $0.25T$ at $fT = 0$ and ± 1, respectively, and equal to zero at other discrete frequencies (in the figure, amplitude is normalized to $W_n(0)$). Therefore, if a sinusoidal signal with $f_0 = m/T$ (m integer) is analysed by use of this window, nonzero components appear at $fT = m - 1$, m, and $m + 1$ with the above amplitudes. Power reduction by the Hanning window is 4.26 dB, which is obtained by integrating $w_n^2(t)$ over $-T/2 \leqslant t \leqslant T/2$ or by $10 \log_{10}(2 \times 0.25^2 + 0.5^2)$. Even if this reduction is accounted for, the measured largest amplitude at $fT = m$ is 1.74 dB lower than the true value since $20 \log_{10}(0.5) + 4.26 = -1.74$. One practical solution for this problem in the FFT analysis is that 1.74 dB is added to the power spectrum display while keeping (and displaying) the true overall power. Without knowing this, one might be surprised to discover (erroneously, of course) that the addition of the individual power spectra gives an overall power that is 1.74 dB larger than the overall power.

An example of measurement of a sinusoidal wave by use of an FFT analyser is shown in Fig. 2.2.7. Conditions of measurement are the same as in Fig. 2.2.5. If $f_0 = 206.5$ Hz, the amplitudes at 200 Hz and 213 Hz are the same with $0.42T$, which is 1.4 dB lower than $0.5T$ (see Fig. 2.2.11). The Hanning window is not suitable for the analysis of periodic signals and is mostly used for analysing stationary random signals with a sufficient number of averages.

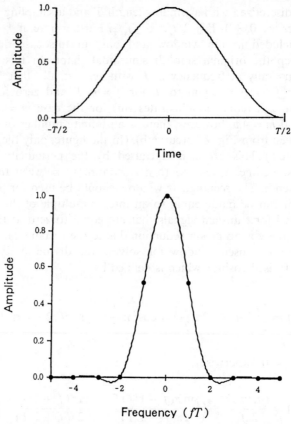

Fig. 2.2.6. The Hanning window (top) and its Fourier transform (bottom).

There are many cases where correct estimation of the amplitudes of a periodic signal is desired. The rectangular window gives correct spectrum amplitudes only when the window length is equal to an integer multiple of the signal period. If the frequency response of the window has a flat region over $-1/T \leqslant f \leqslant 1/T$, this problem is solved. The flat-top window is developed for this purpose. It is defined in the frequency domain as a convolution between the Hamming window and the rectangular filter with a width $-2/T \leqslant f \leqslant 2/T$, where the Hamming window is given by

$$w_m(t) = 0.54 + 0.46 \cos\left(2\pi\frac{t}{T}\right) \tag{2.2.10}$$

Then the flat-top window functions $w_f(t)$, $W_f(f)$ are given, respectively, by

$$w_f(t) = \left(0.54 + 0.46 \cos\left(2\pi\frac{t}{T}\right)\right)\frac{\sin(4\pi t/T)}{4\pi t/T} \tag{2.2.11}$$

and

$$W_f(f) = W_m(f) \otimes W_{2/T}(f) \tag{2.2.12}$$

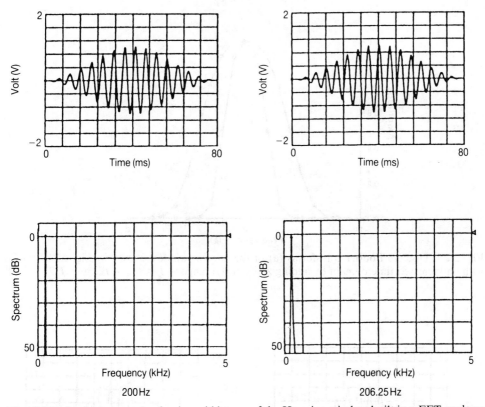

Fig. 2.2.7. Analysis example of a sinusoid by use of the Hanning window built in a FFT analyser.

where $W_{2/T}(f)$ is the rectangular filter. The time window function of the flat-top window is shown in Fig. 2.2.8. The time window has negative regions at both sides, which causes a significant reduction of power (7 dB). The frequency response function can be obtained analytically but can also be calculated by use of the FFT. Results are shown in Fig. 2.2.9. Curve (a) is obtained by FFT from $-T/2$ to $T/2$ with $N = 1024$.

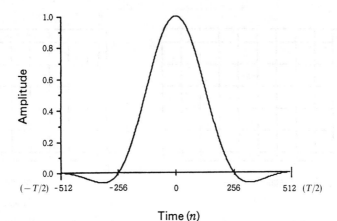

Fig. 2.2.8. The flat-top window.

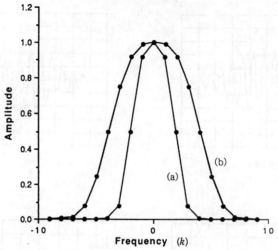

Fig. 2.2.9. Discrete Fourier transforms (a) of the flat-top window shown in Fig. 2.2.8 and (b) with an increased time domain by adding zeros in the range $-T \leqslant t \leqslant -T/2$ and $T/2 \leqslant t \leqslant T$.

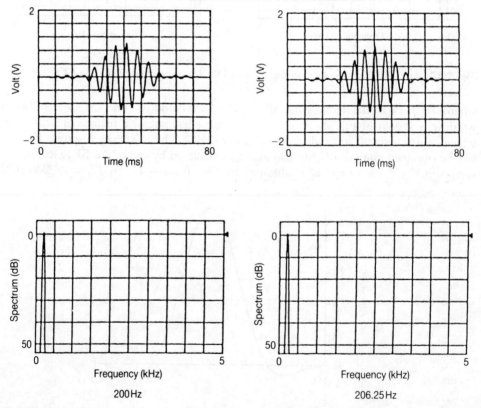

Fig. 2.2.10. Analysis example of a sinusoid by use of the flat-top window built in an FFT analyser.

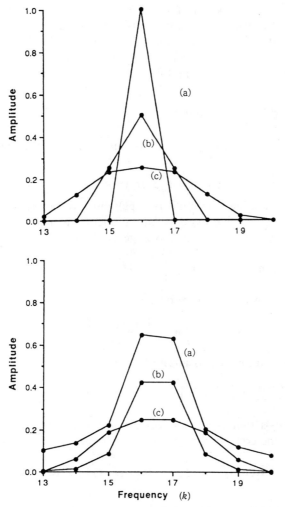

Fig. 2.2.11. Analysis of sinusoidal signals for $fT = 16$ (top) and 16.5 (bottom) by use of the rectangular (a), Hanning (b) and flat-top (c) windows.

The result is shown only from $-10/T$ to $10/T$. It can be seen that the value at $\pm 1/T$ (0.92) is very close to the value at zero frequency. Since we are more interested in the values at $\pm 1/2T$, the range of the time domain is increased to $-T \leqslant t \leqslant T$ by adding zeros between $-T \leqslant t \leqslant -T/2$ and $T/2 \leqslant t \leqslant T$. The number of samplings N is kept the same (1024). The result is shown by curve (b) in Fig. 2.2.9. $W_f(\pm 1)$ is equal to $0.984 W_f(0)$, indicating that the amplitude of a sinusoid can be estimated with a maximum error of -0.15 dB. The advantage of the flat-top window is that components of a periodic signal are very accurately estimated. An example of FFT analysis by use of the flat-top window is shown in Fig. 2.2.10. Even if $f_0 = 206.5$ Hz, the amplitude is accurately determined.

Differences among the rectangular (a), Hanning (b), and flat-top (c) windows are compared in Fig. 2.2.11. Sinusoidal waves with $f_0 T = 16$ (top) and 16.5 (bottom) are digitized with $N = 1024$. Amplitudes are normalized to the value obtained by use of the rectangular window (a) and at $k = 16$ for $f_0 T = 16$. The reduction of the level due to the Hanning and flat-top windows is obvious. The rectangular window has the finest resolution but gives the largest amplitude measurement error if a discontinuity is involved. The flat-top window gives a correct amplitude estimation even for $f_0 T = 16.5$. Various kinds of windows are discussed by Harris (1978).

2.2.3. Correlation and Power Spectrum

The cross correlation function $\Phi_{xy}(t, \tau)$ of two functions $x(t)$ and $y(t)$ is defined as follows:

$$\Phi_{xy}(t, \tau) = E[x^*(t) y(t + \tau)] \tag{2.2.13}$$

where E is the expected value (ensemble average) and τ is the lag. If the process is a stationary random process, the cross correlation function is independent of t and the ensemble average is replaced by the time average as

$$\Phi_{xy}(\tau) = \lim_{T \to \infty} \frac{1}{T} \int_{-T/2}^{T/2} x^*(t) y(t + \tau) \, dt \tag{2.2.14}$$

If we define the cross (power) spectrum by

$$S_{xy}(f) = \lim_{T \to \infty} E\left[\frac{X^*(f) Y(f)}{T} \right] \tag{2.2.15}$$

then the cross correlation function is given by (Thomas 1969)

$$\Phi_{xy}(\tau) = \int_{-\infty}^{\infty} S_{xy}(f) e^{j2\pi f \tau} \, df \tag{2.2.16}$$

Equation (2.2.16) shows that the cross correlation function is given as the inverse Fourier transform of the cross spectrum. The term 'cross' can be replaced by 'auto' if $x(t) = y(t)$.

The cross spectrum given by Eq. (2.2.15) is estimated by use of a two-channel FFT analyser as

$$\langle S_{xy}(k) \rangle_M = \frac{1}{M} \sum_{m=1}^{M} [X_m^*(k) Y_m(k)] \tag{2.2.17}$$

where $\langle \ \rangle_M$ indicates M averagings, and $X_m(k)$ and $Y_m(k)$ are DFTs of the mth sample sets $x_m(n)$ and $y_m(n)$, $n = 0, 1, 2, \ldots, N$, respectively. The number of averagings M is chosen so that Eq. (2.2.17) gives the estimation with adequate accuracy. If $x(t)$ and $y(t)$ are the same, the above equation gives the estimated auto spectrum. The corresponding cross correlation and autocorrelation can be obtained by the inverse DFTs of the cross and auto spectra, respectively.

2.2.4. Frequency Response Function

The output $y(t)$ of a linear system with the impulse response $h(t)$ and input $x(t)$ is given by (see Fig. 2.2.12)

$$y(t) = \int_{-\infty}^{t} x(\tau) h(t - \tau) \, d\tau \tag{2.2.18}$$

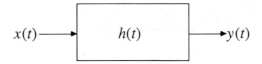

Fig. 2.2.12. A linear system with an impulse response $h(t)$.

This is because the input $x(\tau)\,d\tau$ at time τ contributes to the output by the amount $x(\tau)h(t - \tau)\,d\tau$ at time t. Since $h(t - \tau)$ is zero for $(t - \tau) < 0$, the integration can be extended to $+\infty$:

$$y(t) = \int_{-\infty}^{\infty} x(\tau)h(t - \tau)\,d\tau \tag{2.2.19}$$

This can be rewritten by replacing $(t - \tau)$ by τ as

$$y(t) = \int_{-\infty}^{\infty} x(t - \tau)h(\tau)\,d\tau \tag{2.2.20}$$

This is a convolution of $x(t)$ and $h(t)$. If $x(t)$ and $h(t)$ are absolutely integrable, the Fourier transform of $y(t)$ is given by (see Table 2.1.1)

$$Y(f) = X(f)H(f) \tag{2.2.21}$$

where $X(f)$ and $H(f)$ are the Fourier transforms of $x(t)$ and $h(t)$, respectively. The above equation can be rewritten as

$$H_1(f) = X^*(f)Y(f)/|X(f)|^2 \tag{2.2.22}$$

or

$$H_2(f) = |Y(f)|^2/X(f)Y^*(f) \tag{2.2.23}$$

Equations (2.2.22) and (2.2.23) give the same result if the system is linear and no noise is included in the signals $x(t)$ and $y(t)$.

Discrete forms for stationary signals corresponding to these equations are given, respectively, as

$$\langle H_1(k)\rangle_M = \frac{\dfrac{1}{M}\sum_{m=1}^{M}[X_m^*(k)Y_m(k)]}{\dfrac{1}{M}\sum_{m=1}^{M}[X_m^*(k)X_m(k)]}$$

$$= \frac{\langle S_{xy}(k)\rangle_M}{\langle S_{xx}(k)\rangle_M} \tag{2.2.24}$$

or

$$\langle H_2(k)\rangle_M = \frac{\langle S_{yy}(k)\rangle_M}{\langle S_{yx}(k)\rangle_M} \tag{2.2.25}$$

Equation (2.2.24) is used for most cases since the output (response) tends to include more noise than the input.

2.2.5. Coherence

The coherence function of two signals $x(t)$ and $y(t)$ is defined as

$$\gamma_{xy}(f) = \frac{|S_{xy}(f)|^2}{S_{xx}(f)S_{yy}(f)} \geq 0 \qquad (2.2.26)$$

If $x(t)$ and $y(t)$ are the input and output of a noiseless linear system, $\gamma_{xy}(f)$ is equal to unity since

$$\gamma_{xy}(f) = \frac{S_{xy}(f)/S_{xx}(f)}{S_{yy}(f)/S_{xy}^*(f)} = \frac{H_1(f)}{H_2(f)} = 1 \qquad (2.2.27)$$

(a)

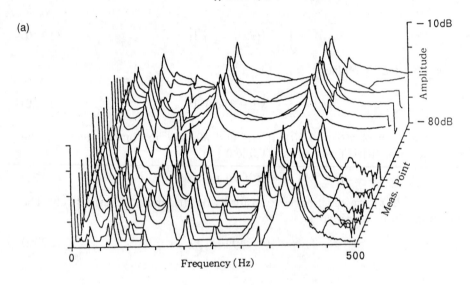

(b)

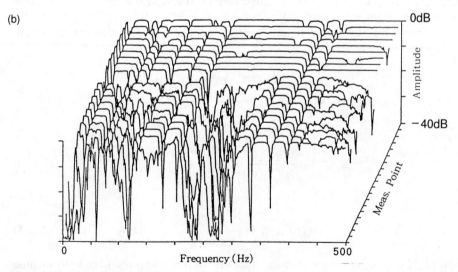

Fig. 2.2.13. Examples of (a) frequency response function and (b) coherence function measurement by use of an FFT analyser.

If noise is included in the input, the measured value of $H_1(f)$ becomes smaller than the true value. On the other hand, if noise is included in the output, the measured value of $H_2(f)$ becomes larger than the true value. Therefore, $\gamma_{xy}(f)$ becomes small whenever noise is involved in the measurement. If the system is nonlinear, $\gamma_{xy}(f)$ also becomes small since the output contains components that are not correlated to the input.

A discrete form of the coherence function is given by

$$\langle \gamma_{xy}(k) \rangle_M = \frac{|\langle S_{xy}(k) \rangle_M|^2}{\langle S_{xx}(k) \rangle_M \langle S_{yy}(k) \rangle_M} \tag{2.2.28}$$

The coherence defined by Eq. (2.2.28) is less than or equal to unity. If the number of averagings is equal to 1, $\langle \gamma_{xy}(k) \rangle_M$ is also equal to unity without regard to the existence of noise, since

$$|\langle S_{xy}(k) \rangle_1|^2 = |X_1^*(k)Y_1(k)|^2 = \langle S_{xx}(k) \rangle_1 \langle S_{yy}(k) \rangle_1 \tag{2.2.29}$$

Another case for $\langle \gamma_{xy}(k) \rangle_M = 1$ is when a periodic input signal is used for a noiseless system. The coherence is less than unity for the following cases:

(1) Noise is included in the system.
(2) The system is nonlinear.
(3) The time window is not long enough compared to the impulse response of the system.

In the third case, the window length must be made longer. Figure 2.2.13 shows examples of frequency response function and coherence measurements. The top chart shows frequency response functions measured on a beam at 20 different points (the excitation point is fixed). The bottom chart shows the corresponding 20 coherence functions. It is seen that the coherence function becomes small at resonance frequencies of the system (due to case (3)). The coherence is also small at frequencies where the frequency response function is small. The reduction of coherence in this case is due to low signal-to-noise ratios.

2.3. Hilbert Transform

The Hilbert transform is widely used in signal processing (Papoulis, 1962; Thomas, 1969). If a signal is Hilbert transformed, phases of positive and negative frequency components are shifted by $\pi/2$ and $-\pi/2$, respectively. By use of this property, an analytic signal that has zero negative frequency components can be obtained. In this section, the causality and analyticity of functions are defined, and then the Hilbert transform and its applications are described.

2.3.1. Sign (or Signum) Function

The sign or signum function is defined by

$$\mathrm{sgn}(t) = 1 \quad \text{for } t \geqslant 0 \quad \text{and} \quad -1 \text{ for } t < 0 \tag{2.3.1}$$

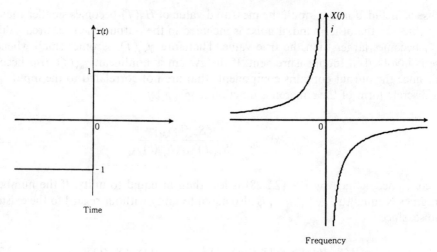

Fig. 2.3.1. Fourier transform pair of a sign (signum) function.

Its Fourier transform is given by

$$X(f) = \int_{-\infty}^{\infty} \text{sgn}(t) e^{-j2\pi ft}\, dt = -j \int_{0}^{\infty} 2\, \sin(2\pi ft)\, dt$$

$$= \frac{1}{j\pi f} \tag{2.3.2}$$

The transform pair of the sign function is shown in Fig. 2.3.1 (same as Fig. 2.1.13). The above integration must be interpreted as a distribution (Papoulis, 1962; Weaver, 1989). A validity of Eq. (2.3.2) is confirmed by the inverse Fourier transform of $X(f)$ (Gradshteyn and Ryzhik, 1965).

$$x(t) = \int_{-\infty}^{\infty} \frac{1}{j\pi f}\, e^{j2\pi ft}\, df = \frac{1}{\pi} \int_{-\infty}^{\infty} \frac{\sin 2\pi ft}{2\pi f}\, d(2\pi f)$$

$$= \text{sgn}(t) \tag{2.3.3}$$

In the same way, for a function

$$x(t) = \frac{-1}{\pi t} \tag{2.3.4}$$

the Fourier transform is given by

$$X(f) = -\int_{-\infty}^{\infty} \frac{1}{\pi t}\, e^{-j2\pi ft}\, dt = j\, \text{sgn}(f) \tag{2.3.5}$$

2.3.2. Hilbert Transform

The relation between a function $x(t)$ and its Hilbert transform $x_h(t)$ is given by

$$x_h(t) = -\frac{1}{\pi} \int_{-\infty}^{\infty} \frac{x(t')}{t - t'}\, dt' \tag{2.3.6}$$

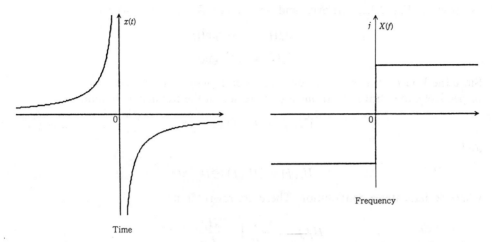

Fig. 2.3.2. Impulse response and frequency response functions of a Hilbert transformer.

and

$$x(t) = \frac{1}{\pi} \int_{-\infty}^{\infty} \frac{x_h(t')}{t - t'} \, dt' \qquad (2.3.7)$$

Equation (2.3.6) indicates that the Hilbert transform of $x(t)$ is given as a convolution between $x(t)$ and $-1/\pi t$. If $x(t) = \delta(t)$, then $x_h(t) = -1/\pi t$. Therefore, the impulse response of a Hilbert transformer is considered to be

$$h(t) = \frac{-1}{\pi t} \qquad (2.3.8)$$

Its Fourier transform is given by Eq. (2.3.5). The impulse response and the frequency response function of the Hilbert transformer (Fourier transform pair) are shown in Fig. 2.3.2. The magnitude of the frequency response function is equal to unity and the phase shift is $+\pi/2$ for positive frequencies and $-\pi/2$ for negative frequencies.

2.3.3. Causality

If the impulse response of a system $h(t)$ has the property

$$h(t) = 0 \quad \text{for } t < 0 \qquad (2.3.9)$$

the system is causal, that is, the system does not respond before the input.

An impulse response of a physically realizable system is causal. The real and imaginary parts of a frequency response function of this kind of system are related to each other by the Hilbert transform. If the real part is given, the imaginary part is determined from the real part and vice versa. Let us express the frequency response function of a causal system by

$$H(f) = H_r(f) + j H_i(f) \qquad (2.3.10)$$

The impulse response can be expressed as a sum of even ($h_e(t)$) and odd ($h_o(t)$) functions as shown in Fig. 2.3.3. The even and odd functions are related to each other by

$$h_o(t) = h_e(t) \, \text{sgn}(t) \qquad (2.3.11a)$$

$$h_e(t) = h_o(t) \, \text{sgn}(t) \qquad (2.3.11b)$$

Since the Fourier transforms of even and odd functions are purely real and imaginary, respectively, the above equations are expressed in the frequency domain by

$$jH_i(f) = H_r(f) \otimes (1/j\pi f) \qquad (2.3.12)$$

and

$$H_r(f) = jH_i(f) \otimes (1/j\pi f) \qquad (2.3.13)$$

where $\otimes$ indicates a convolution. These are rewritten as

$$H_i(f) = -\frac{1}{\pi} \int_{-\infty}^{\infty} \frac{H_r(f')}{f - f'} \, df' \qquad (2.3.14)$$

$$H_r(f) = \frac{1}{\pi} \int_{-\infty}^{\infty} \frac{H_i(f')}{f - f'} \, df' \qquad (2.3.15)$$

If the impulse response contains an impulse at $t = 0$, Eq. (2.3.15) must be modified as (Papoulis, 1962)

$$H_r(f) = H_r(\infty) + \frac{1}{\pi} \int_{-\infty}^{\infty} \frac{H_i(f')}{f - f'} \, df'. \qquad (2.3.16)$$

2.3.4. Minimum Phase Shift Property

Even if two frequency response functions have the same amplitude characteristic, their phase characteristics may be different. The minimum phase shift property is a property

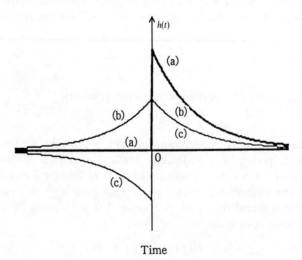

Fig. 2.3.3. A causal function (a), and its representation by even (b) and odd (c) functions.

of a frequency response function that has the minimum phase characteristic for a given amplitude characteristic. For example,

$$H_1(f) = \frac{1}{jf + 2} \tag{2.3.17}$$

and

$$H_2(f) = \frac{1}{jf + 2} \frac{jf - 1}{jf + 1} \tag{2.3.18}$$

have the same amplitude characteristic

$$|H(f)| = \frac{1}{\sqrt{f^2 + 4}} \tag{2.3.19}$$

However, $\arg(H_1(f))$ and $\arg(H_2(f))$ are different:

$$\arg(H_1(f)) = \arg(-jf + 2) = \phi_1 \tag{2.3.20}$$

and

$$\arg(H_2(f)) = \phi_1 + \arg(2jf + f^2 - 1) = \phi_1 + \phi_2 \tag{2.3.21}$$

Vector representations of $(2 - jf)$ and $(f^2 - 1 + 2jf)$ are shown in Fig. 2.3.4 at various frequencies from $-\infty$ to $+\infty$. The phase characteristics of $H_1(f)$ and $H_2(f)$ are shown in Fig. 2.3.5. As the figures show, $H_2(f)$ has a larger phase shift than $H_1(f)$.

For a function with the minimum phase shift property, the amplitude response is determined from the phase response and vice versa (Papoulis, 1962). If a frequency response function $H(f)$ with the minimum phase shift property is expressed by

$$H(f) = e^{-\alpha(f) - j\phi(f)} \tag{2.3.22}$$

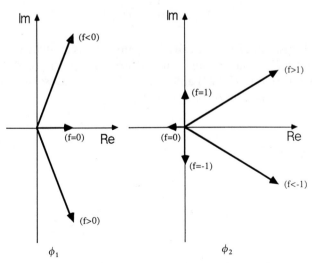

Fig. 2.3.4. Vector representations of ϕ_1 and ϕ_2.

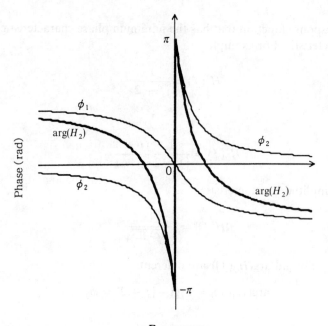

Fig. 2.3.5. Phase characteristics of minimum ($H_1(f)$) and nonminimum ($H_2(f)$) phase functions.

the amplitude and phase characteristics satisfy the following relationships:

$$\phi(f) = -\frac{f}{\pi} \int_{-\infty}^{\infty} \frac{\alpha(f')}{f^2 - f'^2} \, df' \tag{2.3.23}$$

and

$$\alpha(f) = \alpha(0) + \frac{f^2}{\pi} \int_{-\infty}^{\infty} \frac{\phi(f')}{f'(f^2 - f'^2)} \, df' \tag{2.3.24}$$

If a Laplace transform $H(p)$ of a causal function $h(t)$ is analytic and has no zeros in the right half-plane, $H(f)$ has the minimum phase shift property and its amplitude and phase functions satisfy Eqs. (2.3.23) and (2.3.24). This is because $\alpha(f)$ and $\phi(f)$ are the real and imaginary parts of $\ln H(f)$, respectively.

2.3.5. Analytic Signal

Since the Laplace transform, $H(p)$, of a causal function ($h(t) = 0$ for $t < 0$) is analytic on the right-half-plane with the complex variable $p = \alpha + j2\pi f$, a signal that has zero spectrum for $f < 0$ is conventionally called as an 'analytic signal'.

An analytic signal, $x_a(t)$, is obtained from the original signal, $x(t)$, by

$$x_a(t) = x(t) - jx_h(t) \tag{2.3.25}$$

The Fourier transform of this signal is

$$
\begin{aligned}
X_a(f) &= X(f) - j[j\operatorname{sgn}(f)X(f)] \\
&= X(f) + \operatorname{sgn}(f)X(f) \\
&= 2X(f) \qquad \text{for } f \geqslant 0 \\
&\quad\ 0 \qquad\qquad \text{for } f < 0
\end{aligned}
\tag{2.3.26}
$$

For the analytic signal to have the same energy as the original signal, the right-hand side of Eq. (2.3.25) must be divided by $\sqrt{2}$ (Thomas, 1969).

Since the Hilbert transform of the impulse $\delta(t)$ is $-1/\pi t$, the analytic impulse is

$$
\delta_a(t) = \delta(t) + \frac{j}{\pi t}
\tag{2.3.27}
$$

The analytic signal $x_a(t)$ is considered as an output of a system with an impulse response $\delta_a(t)$ and input $x(t)$. The most simple example of the analytic signals is that of the cosine or sine function. Let $x(t) = \cos 2\pi ft$, then $x_h(t) = -\sin 2\pi ft$. Therefore, we have $x_a(t) = e^{j2\pi ft}$. Similarly, in case of $f(t) = \sin 2\pi ft$, $x_a(t) = -je^{j2\pi ft}$.

In the application of the Wigner distribution, a signal is made analytic in order to avoid the interference between the positive and negative frequency components (see Section 2.4.3).

2.3.6. Instantaneous Envelope and Phase

When an analytic signal $x_a(t)$ is represented by amplitude and phase functions as

$$
x_a(t) = A(t)e^{j\phi(t)}
\tag{2.3.28}
$$

where $A(t)$ and $\phi(t)$ are real functions. The instantaneous envelope and phase are given, respectively, by

$$
|A(t)|^2 = x^2(t) + x_h^2(t)
\tag{2.3.29}
$$

$$
\phi(t) = \tan^{-1}[x_h(t)/x(t)]
\tag{2.3.30}
$$

The instantaneous frequency is obtained by differentiating $\phi(t)$ with respect to t as

$$
f_i(t) = \frac{1}{2\pi}\frac{\partial \phi(t)}{\partial t}
\tag{2.3.31}
$$

Example. Consider a cosine function with frequency f_c, whose amplitude is modulated by a sine function with bandwidth $2f_m$:

$$
x(t) = \frac{\sin 2\pi f_m t}{2\pi f_m t} \cos 2\pi f_c t \qquad (f_m < f_c)
\tag{2.3.32}
$$

If the condition $f_m < f_c$ is satisfied, the analytic function is simply given as

$$
x_a(t) = \frac{\sin 2\pi f_m t}{2\pi f_m t} e^{j2\pi f_c t}
\tag{2.3.33}
$$

and the envelope function does not change.

Now let us obtain the instantaneous envelope and phase of the sinc function itself. Let

$$x(t) = \frac{\sin 2\pi f_m t}{2\pi f_m t} \tag{2.3.34}$$

then (see Gradshteyn and Ryzhik, 1965)

$$x_h(t) = -2\frac{\sin^2 \pi f_m t}{2\pi f_m t} \tag{2.3.35}$$

Therefore,

$$x_a(t) = \frac{\sin \pi f_m t}{\pi f_m t}\, e^{j\pi f_m t} \tag{2.3.36}$$

The envelope function is still a sinc function and its frequency is 1/2 of the original sinc function (see Fig. 2.3.6). The phase function is

$$\phi(t) = \pi f_m t \tag{2.3.37}$$

and the instantaneous frequency is

$$f_i(t) = f_m/2 \tag{2.3.38}$$

which is independent of time.

It is sometimes useful to look at a signal not only by its time waveform or frequency spectrum but also by its instantaneous envelope and phase or frequency (Suzuki, 1980).

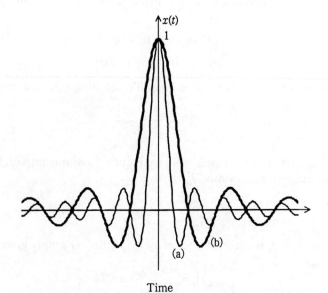

Fig. **2.3.6.** Sinc function (a) and its envelope function (b).

2.4. Time–Frequency Domain Analysis

Signals we observe in the real world are functions mostly of time or space. When the properties of the signal are analysed in the time or the space domain, statistical values such as the probability density function, average, deviation, or even higher-order moments are used. On the other hand, when the spectrum of the signal is required, it is Fourier transformed and the properties in the frequency domain are discussed. Since the real signal is recovered by the inverse Fourier transform of the Fourier spectrum, it is considered that the real signal and the spectrum (mostly complex-valued) are two ways of representing the same phenomenon. In other words, we are observing the same phenomenon from the front or from the side (or rear). There are some occasions, however, when we wish to be able to represent one phenomenon in the 2-dimensional time and frequency domain. For example, when a musical instrument is played, a 'time window' can be used to separate one note from others, and the Fourier spectrum of the specific note is obtained. If transient properties of a musical instrument sound are analysed, a very short time window can be used to separate the onset, sustain and decay of the signal. The spectrogram is most commonly used for this kind of purpose. In this section, several ways of representing the signal in the time–frequency domain will be introduced.

2.4.1. Spectrogram

The spectrogram is a method of looking at a signal through a (short) window and to obtain the Fourier spectrum of the windowed signal (Rabiner and Shafer, 1978). If the position of the window is moved, we can observe the gradual change of the signal spectrum as a function of time. Let the signal and the window functions be $x(t)$ and $w(t)$, respectively. Then the windowed signal at a fixed time t can be represented as a function of time τ by $x(\tau)w(\tau - t)$. The spectrum of this windowed signal, which is a function of time t and frequency f, is given by

$$X(t, f) = \int_{-\infty}^{\infty} x(\tau)w(\tau - t)e^{-j2\pi f\tau} d\tau \qquad (2.4.1)$$

The most common window is a raised cosine (Hanning window), which is defined by

$$w(t) = \cos^2\left(\frac{\pi t}{T}\right) \qquad -T \leqslant t \leqslant T$$

$$0 \qquad\qquad \text{otherwise} \qquad (2.4.2)$$

A disadvantage of the spectrum is that the frequency resolution is degraded if the width of the window is made narrower in order to increase the time resolution.

2.4.2. Page's, Levin's, and Rihaczek's Methods

Consider that a time function $x(t)$ is comprised of $x_1(t)$ and $x_2(t)$ such that

$$x_1(t, \tau) = x(\tau), \; x_2(t, \tau) = 0 \qquad -\infty \leqslant \tau \leqslant t \qquad (2.4.3)$$

$$x_2(t, \tau) = x(\tau), \; x_1(t, \tau) = 0 \qquad t \leqslant \tau \leqslant \infty \qquad (2.4.4)$$

The energy spectrum of $x_1(t, \tau)$ and $x_2(t, \tau)$ are given, respectively, by

$$E_1(t, f) = |X_1(t, f)|^2 \qquad (2.4.5)$$

and

$$E_2(t, f) = |X_2(t, f)|^2 \qquad (2.4.6)$$

where $X_1(t, f)$ and $X_2(t, f)$ are the Fourier transforms of $x_1(t, \tau)$ and $x_2(t, \tau)$, respectively. Then the power spectrum will be obtained by differentiating them with respect to t:

$$P_1(t, f) = \frac{\partial}{\partial t} E_1(t, f) \qquad (2.4.7)$$

and

$$P_2(t, f) = -\frac{\partial}{\partial t} E_2(t, f) \qquad (2.4.8)$$

The negative sign in Eq. (2.4.8) is necessary because $\int_{-\infty}^{+\infty} E_2(t, f)\, df$ is a monotonously decreasing function. $P_1(t, f)$ and $P_2(t, f)$ are the energy density distributions of $f(t)$ on the time–frequency $(t-f)$ plane proposed by Page (1952) and Levin (1964), respectively. Since $X_1(t, f)$ and its derivative are given by

$$X_1(t, f) = \int_{-\infty}^{\infty} x_1(t, \tau)e^{-j2\pi f \tau}\, d\tau = \int_{-\infty}^{t} x(\tau)e^{-j2\pi f \tau}\, d\tau \qquad (2.4.9)$$

and

$$\frac{\partial}{\partial t}(X_1(t, f)) = x(t)e^{-j2\pi f t} \qquad (2.4.10)$$

respectively, Eq. (2.4.7) can be rewritten as

$$P_1(t, f) = \frac{\partial}{\partial t} |X_1(t, f)|^2$$

$$= \frac{\partial}{\partial t}(X_1(t, f))X_1^*(t, f) + X_1(t, f)\frac{\partial}{\partial t}(X_1^*(t, f))$$

$$= x(t)e^{-j2\pi f t}\int_{-\infty}^{t} x(\tau)\, e^{j2\pi f \tau}\, d\tau + x(t)e^{j2\pi f t}\int_{-\infty}^{t} x(\tau)\, e^{-j2\pi f \tau}\, d\tau$$

$$= 2x(t)\,\mathrm{Re}[e^{j2\pi f t}X_1(t, f)] \qquad (2.4.11)$$

Similarly,

$$P_2(t, f) = 2x(t)\,\mathrm{Re}[e^{j2\pi f t}X_2(t, f)] \qquad (2.4.12)$$

$P_1(t, f)$ and $P_2(t, f)$ are the instantaneous power spectra obtained by use of information only in the past and future, respectively. Rihaczek (1968) proposed a method of using the whole information of the signal from the past to the future, that is

$$P_3(t, f) = \tfrac{1}{2}[P_1(t, f) + P_2(t, f)]$$

$$= x(t)\,\mathrm{Re}[e^{j2\pi f t}X(f)] \qquad (2.4.13)$$

Example 1. If $x(t) = A \sin 2\pi f_0 t$,

$$X(f) = \frac{A}{2j}[\delta(f - f_0) - \delta(f + f_0)]$$

and, therefore,

$$P_3(t, f) = A \sin 2\pi f_0 t \, (\sin 2\pi ft) \, \frac{A}{2} [\delta(f - f_0) - \delta(f + f_0)]$$

$$= (A \sin 2\pi f_0 t)^2 / 2 \qquad (f = \pm f_0)$$

$$0 \qquad\qquad (f \neq \pm f_0) \qquad\qquad (2.4.14)$$

which is equal to the normally defined instantaneous power of $x(t)$, whose spectrum exists only at $\pm f_0$.

Example 2. If $x(t) = A(\sin 2\pi f_0 t)/\pi t$ (sinc function)

$$X(f) = A \qquad -f_0 < f < f_0$$

$$0 \qquad \text{otherwise}$$

and, therefore,

$$P_3(t, f) = A^2 (\sin 2\pi f_0 t)(\cos 2\pi ft)/\pi t \qquad -f_0 < f < f_0$$

$$0 \qquad\qquad \text{otherwise} \qquad\qquad (2.4.15)$$

In the case $A = 1$ and $f_0 = 1$ (Hz), $P_3(t, f)$ at fixed times ($t = 0.0, 0.1, 0.2, 0.4, 0.8$, and 1.6) in the region $0 < f < 1$ is shown in Fig. 2.4.1. $P_3(t, f)$ is symmetric for both time and frequency. In some regions on the $t-f$ plane, the distribution becomes negative.

A common property of the above three methods is that, at times when $x(t) = 0$, the distributions on the $t-f$ plane are also zero at those times. This requirement seems to be too strict for real application of these methods.

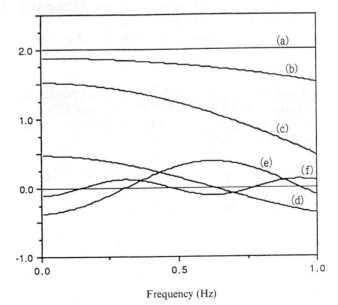

Frequency (Hz)

Fig. 2.4.1. Rihaczek distribution of a sinc function $(\sin 2\pi t)/\pi t$ at $t = 0$ (a), 0.1 (b), 0.2 (c), 0.4 (d), 0.8 (e) and 1.6 s (f).

2.4.3. Wigner Distribution

The autocorrelation and the auto spectrum are a Fourier transform pair of each other defined for stationary signals (see Eq. (2.2.13)).

$$\Phi_{xx}(\tau) = \mathrm{E}[x^*(t)x(t+\tau)] \leftrightarrow S_{xx}(f) = \int_{-\infty}^{\infty} \Phi_{xx}(\tau)e^{-j2\pi f\tau}\,d\tau \qquad (2.4.16)$$

For a nonstationary signal, which must be dealt with as a deterministic signal, the autocorrelation can be defined as a function of both time t and the time difference (lag) τ. In general,

$$\Phi_{xx}(t, \tau) = x^*(t+k\tau)x(t+(k+1)\tau) \qquad (2.4.17)$$

where * indicates the complex conjugate, which is not necessary if the signal is real (Hino, 1983). If the autocorrelation is defined with $k = 0$ in the above equation, the corresponding distribution on the $t-f$ plane is given by

$$\int_{-\infty}^{\infty} x^*(t)x(t+\tau)e^{-j2\pi f\tau}\,d\tau = x^*(t)\int_{-\infty}^{\infty} x(\tau')e^{-j2\pi f(\tau'-t)}\,d\tau'$$
$$= x^*(t)e^{j2\pi ft}X(f) \qquad (2.4.18)$$

For $k = -1$,

$$\int_{-\infty}^{\infty} x^*(t-\tau)x(t)e^{-j2\pi f\tau}\,d\tau = x(t)e^{-j2\pi ft}X(-f) \qquad (2.4.19)$$

If we assume that $x(t)$ is real, that is, $x(t) = x^*(t)$ and $X(-f) = X^*(f)$, then the average of Eqs (2.4.18) and (2.4.19) gives the distribution proposed by Rihaczek (1968),

$$P_3(t, f) = x(t)\,\mathrm{Re}[e^{j2\pi ft}X(f)] \qquad (2.4.20)$$

Equation (2.4.13) was derived from the physical point of view. On the other hand, Eq. (2.4.20) was obtained by the natural extension of the relation (Eq. (2.4.16)) for the stationary case to the nonstationary (deterministic) case.

In order to make the autocorrelation symmetric with respect to τ, let k equal $-1/2$ in Eq. (2.4.17).

$$\Phi_{xx}(t, \tau) = x^*(t-\tau/2)x(t+\tau/2) \qquad (2.4.21)$$

The power spectrum obtained by the Fourier transform of this autocorrelation is called the auto Wigner distribution (Wigner, 1932; Claasen and Mecklenbräuker, 1980), or the Wigner–Ville distribution (Cohen, 1989), or the generalized spectral density (Bendat and Piersol, 1971):

$$W(t, f) = \int_{-\infty}^{\infty} x^*(t-\tau/2)x(t+\tau/2)e^{-j2\pi f\tau}\,d\tau \qquad (2.4.22)$$

The Wigner distribution can also be defined by use of the Fourier spectrum, such as

$$W(f, t) = \int_{-\infty}^{\infty} X^*(f-\zeta/2)X(f+\zeta/2)e^{j2\pi \zeta t}\,d\zeta \qquad (2.4.23)$$

The Wigner distribution has interesting properties such as the following.

(1) Integration with respect to frequency gives the instantaneous power.

$$\int_{-\infty}^{\infty} W(t, f)\,df = |x(t)|^2 \qquad (2.4.24)$$

(2) Integration with respect to time gives the energy spectrum.

$$\int_{-\infty}^{\infty} W(t, f)\, dt = |X(f)|^2 \tag{2.3.25}$$

(3) The Wigner distribution of the time-shifted signal is also time-shifted.

$$x(t - \tau) \rightarrow W(t - \tau, f)$$

(4) The Wigner distribution of frequency shifted signal is also frequency shifted.

$$X(f - \zeta) \rightarrow W(t, f - \zeta)$$

(5) The Wigner distribution of the time-limited signal is also time-limited in the same region.

If $x(t) = 0$ for $t < a$ and $t > b$, then $W(t, f) = 0$ for $t < a$ and $t > b$.

(6) The Wigner distribution of the frequency-limited signal is also frequency-limited in the same region.

If $X(f) = 0$ for $f < a$ and $f > b$, $W(t, f) = 0$ for $f < a$ and $f > b$.

(7) The centre of gravity of the Wigner distribution with respect to time is equal to the group delay.

$$\int_{-\infty}^{\infty} tW(t, f)\, dt \Big/ \int_{-\infty}^{\infty} W(t, f)\, dt = t_g(f) \tag{2.4.26}$$

(8) The centre of gravity of the Wigner distribution with respect to frequency is equal to the instantaneous frequency.

$$\int_{-\infty}^{\infty} fW(t, f)\, df \Big/ \int_{-\infty}^{\infty} W(t, f)\, df = f_i(t) \tag{2.4.27}$$

(9) The Wigner distribution takes negative values.

Equations (2.4.24) and (2.4.25) indicate that the Wigner distribution is a good candidate to be interpreted as the energy density distribution on the $t-f$ plane, but the last property contradicts the conventional understanding of energy. It seems meaningless to try to understand the physical meaning of the Wigner distribution (and any other distributions on the $t-f$ plane) in a strict sense since it contravenes the uncertainty principle.

Equation (2.4.22) is rewritten so that the Fourier transform can be applied to the Wigner distribution (Janse and Kaizer, 1983). Replacing $\tau/2$ by τ' and $2f$ by f'

$$W(t, f'/2) = \int_{-\infty}^{\infty} 2x(t + \tau')x(t - \tau')e^{-j2\pi f'\tau'}\, d\tau' \tag{2.4.28}$$

where $x(t)$ is assumed to be real. Letting

$$u_t(\tau') = 2x(t + \tau')x(t - \tau') \tag{2.4.29}$$

and again replacing τ' by τ and f' by f, we obtain

$$W(t, f/2) = \int_{-\infty}^{\infty} u_t(\tau)e^{-j2\pi f\tau}\, d\tau = U_t(f) \tag{2.4.30}$$

where $U_t(f)$ is the Fourier transform of $u_t(\tau)$. Note that the component of the Wigner distribution at f is obtained as the component of $U_t(f)$ at $2f$.

For a practical application of the Wigner distribution, the Wigner distribution must be used in the limited time and frequency range. The discretization of the Wigner distribution is as follows.

Let

$$t = n\Delta T \qquad \tau = m\Delta T \qquad 2\pi f \Delta T = \Omega$$

where ΔT is the sampling period. Then the discrete Wigner distribution is given by

$$W(n, \Omega/2) = \Delta T \sum_{m=-\infty}^{\infty} u_n(m)e^{-jm\Omega} \tag{2.4.31}$$

where

$$u_n(m) = 2x(n + m)x(n - m) \tag{2.4.32}$$

If we use a time window $w(n)$ with a finite duration from $-M/2 + 1$ to $M/2$, then the finite length discrete Wigner distribution for the unit sample period ($\Delta T = 1$) is obtained by

$$W\left(n, \frac{k}{M}\pi\right) = \sum_{m=-M/2+1}^{M/2} u_{n,M}(m)e^{-j2\pi mk/M} \tag{2.4.33}$$

where

$$u_{n,M}(m) = 2x(n + m)w(m)x(n - m)w(-m) \tag{2.4.34}$$

Since $u_{n,M}(m)$ has twice the original frequency range, the windowed signal should be made an analytic signal to avoid the aliasing problem. This also avoids components produced by the interference between positive and negative frequency components (Janse and Kaizer, 1983).

Plate I shows a comparison of the $t-f$ distributions of a chirp (swept sine) signal using the spectrogram (top) and the finite-length discrete Wigner distribution (bottom). The Wigner distribution gives a much finer resolution than the spectrogram in this example. Another example of the application of the Wigner distribution to a vocalization ('cat') is given in Plate II. As this figure shows, the Wigner distribution provides a different way of observing a signal. It should be noted, however, that the Wigner distribution gives useful results for a limited range of signals and applications. Applications of the Wigner distribution to sound intensity analysis is described in Section 4.4.2.

2.4.4. Other Distributions

The $t-f$ distributions described in the previous sections are variations of the wider class of $t-f$ distribution given by Cohen (1966, 1989):

$$C_f(t, \omega; \phi) = \frac{1}{2\pi} \int_{-\infty}^{\infty} \int_{-\infty}^{\infty} \int_{-\infty}^{\infty} e^{j(\xi t - \tau\omega - \xi u)}\phi(\xi, \tau)x^*(u - \tau/2)x(u + \tau/2)\,du\,d\tau\,d\xi \tag{2.4.35}$$

Various kinds of $t-f$ distributions have been proposed, adopting different functions for $\phi(\xi, \tau)$.

The recently developed wavelet transform is one of those distributions (Combes et al., 1989). The wavelet transform of $x(t)$ is given by

$$\tilde{x}(a, b) = \int_{-\infty}^{\infty} \phi_{a,b}^*(t)x(t)\,dt \tag{2.4.36}$$

where $\phi(t)$ is an analysing wavelet that is locally distributed in both time and frequency. After time-scaling and frequency-shifting, $\phi_{a,b}(t)$ is given by

$$\phi_{a,b}(t) = a^{-1}\phi((t-b)/a) \qquad (2.4.37)$$

The wavelet transform does not have the property of symmetry between time and frequency, but it is believed that it can localize a transient phenomenon on the $t-f$ plane better than the spectrogram does. Figure 2.4.2 shows examples of $t-f$ distribution analysis of mixed chirp signals (Terada et al. 1994).

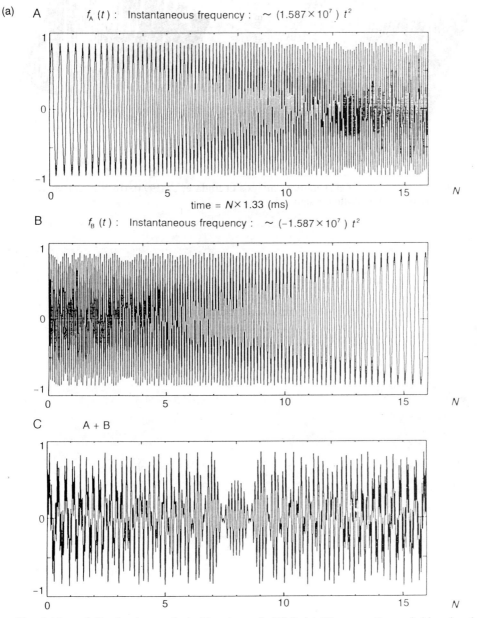

Fig. 2.4.2. $t-f$ distribution analysis (Terada et al. 1994). (a) Time-waveform of chirp signals.

(b)

Kernel Function : $\Phi(m) = \dfrac{\omega_0}{\omega_c}\, e^{\,j\omega_c m}\, e^{\frac{\omega_c}{\omega_0}\frac{1}{2}}$

(c)

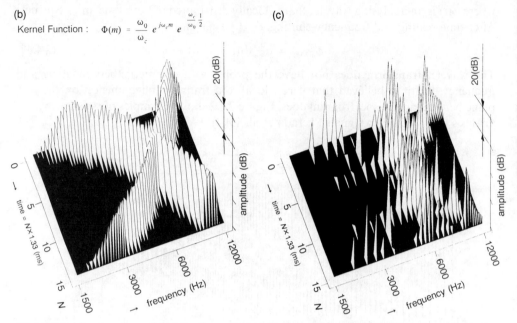

Fig. 2.4.2. (*Continued*). t–f distribution analysis (Terada et al. 1994). (b) Wavelet analysis for signal A + B. (c) Wigner-Ville for signal A + B.

3

Sound Propagation
in an Enclosed Space

This chapter presents a review of fundamental characteristics of sound fields in enclosed spaces. It covers wave phenomena near the walls, resonance and normal modes, transfer functions, modulation transfer functions, random sound fields and reverberation theory, power response of a sound source. These conventional subjects, are very helpful to understanding sound fields. In addition, transfer function phase analysis may open new technologies for sound field control. Lyon (1987) used statistical phase analysis (SPA) in addition to his earlier statistical energy analysis (SEA) (Lyon, 1975). Some new results from use of SPA will be presented in Section 3.5. Applications and new technologies of sound space control will be introduced in Chapters 6 and 7.

3.1. Reflection at a Boundary

Single reflection greatly changes the spatial and frequency transmission characteristics of sound. We call the typical spatial patterns of sound distribution due to reflection 'interference patterns'. Waterhouse (1955) stated theoretically that sound energy is not uniformly distributed, even in a reverberant sound field, but is distributed as 'interference patterns' at the boundaries and at any other reflecting surfaces that are large compared to the wavelength of the sound. This arises from the fact that although the mean energy at all points is the same in all directions, the phases of the wave trains near the reflecting surfaces are not entirely random owing to the reflection of some of the waves by these surfaces. In this chapter, we summarize the fundamental phenomenon caused by single reflection at a boundary in a space based on Waterhouse (1955). Reflected sound also significantly affects our perception. Control of the reflected sound in a concert hall will be discussed in Chapter 7.

3.1.1. Interference Patterns at a Boundary

A sinusoidal interference pattern is produced when plane waves strike a plane reflector. This is due to the addition of two wave trains, the incident and the reflected, with a phase difference between them that varies with the distance from the reflector. At the reflector, the two wave trains are in phase; at a quarter-wavelength distance from the reflector, the wave trains are exactly out of phase. The interference patterns depend on the incidence angles of the waves (see Section 4.2.1).

In Fig. 3.1.1, we consider the plane YOZ to be a perfect reflecting boundary that is large compared to the wavelength. Under this condition, sound energy can reach the point $(x, 0, 0)$ only from the hemisphere to the above of YOZ. Consider a pressure plane wave of unit amplitude incident at angle θ with the x-axis and at angle ϕ with the XOY plane, producing a pressure $\cos \omega t$ at the point $(x, 0, 0)$. It is reflected specularly in the plane YOZ with no phase change, and the reflected wave gives rise to a pressure $\cos(\omega t + 2kx \cos \theta)$ at the point where the wavenumber $k = 2\pi/\lambda$.

The total pressure fluctuation at the point $(x, 0, 0)$ is thus

$$p = \cos(\omega t) + \cos(\omega t + 2kx \cos \theta) \quad (\text{Pa}) \qquad (3.1.1)$$

and the mean squared pressure is

$$\overline{p^2} = 1 + \cos(2kx \cos \theta) \quad (\text{Pa}^2) \qquad (3.1.2)$$

where $^-$ denotes the time average. This interference pattern is plotted in Figs 3.1.2a, 3.1.2b and 3.1.2c for incidence angles $0°$, $30°$, and $60°$, respectively. We can see the periodic peaks and dips in the interference patterns.

An interference pattern is also produced under random incidence conditions. While we normally expect sound pressure to be distributed uniformly under the random incidence conditions, the sound field in the areas close to the boundaries is not uniform. The corresponding expression for $\langle p^2 \rangle$ averaged for waves incident on the plane YOZ

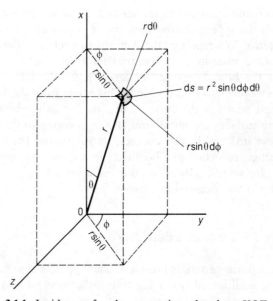

Fig. 3.1.1. Incidence of a plane wave into the plane YOZ.

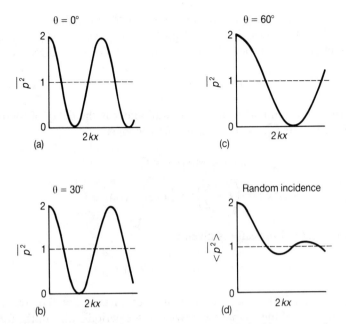

Fig. 3.1.2. Interference patterns produced when sound is incident on a plane reflector at angles 0°, 30°, 60°, and from all directions over a hemisphere (random incidence). x denotes the distance from the reflector. After Waterhouse (1955).

from all directions over a hemisphere is

$$\langle \overline{p^2} \rangle = \frac{1}{2\pi} \int_0^{2\pi} \int_0^{\pi/2} \overline{p^2} \sin\theta \, d\theta \, d\phi$$

$$= 1 + \frac{\sin 2kx}{2kx} \quad (\text{Pa}^2) \tag{3.1.3}$$

where $\sin\theta \, d\theta \, d\phi$ is the solid angle element.

The pattern for random incidence is plotted in Fig. 3.1.2d. We can no longer see clear dips in the pattern. The mean squared pressure at the boundary is just twice that obtained away from the boundary. This relationship is used to estimate the space average of the mean squared pressure for the sound power measurement in a reverberation room (Tohyama and Suzuki, 1986).

3.1.2. Frequency Characteristics of Direct and Single-reflection Sound

Reflection of sound produces not only a typical spatial distribution but also frequency characteristics. The frequency characteristics of a sound path that includes both direct and single-reflection sound is a typical example of frequency response in a sound space.

Again consider a sound field that includes a solid wall. Suppose that the direct sound observed at point M is $A\cos(\omega t)$. The sound pressure observed at M is expressed as the sum of the direct sound and the reflected sound that follows the direct wave:

$$p = A[\cos(\omega t) + R\cos(\omega(t - \tau))] \quad \text{(Pa)} \tag{3.1.4}$$

where R denotes the reflection coefficient of the wall and τ is the delay of the reflected sound. If we set $R = 1$, sound pressure at point M becomes

$$p = A[\cos(\omega t) + \cos(\omega(t - \tau))]$$

$$= 2A\cos\omega\left(t - \frac{\tau}{2}\right)\cos\left(\omega\frac{\tau}{2}\right) \quad \text{(Pa)} \tag{3.1.5}$$

Consequently, mean squared pressure is given as

$$\overline{p^2} = A^2(1 + \cos\omega\tau) \quad \text{(Pa}^2\text{)} \tag{3.1.6}$$

This is plotted in Fig. 3.1.3. The squared pressure response is not flat, but changes according to the frequency of the sound wave. We can see periodic dips in the frequency characteristic. These dips are due to the 'zeros' of the response. That is, mean squared pressure becomes zero at the frequencies where $\omega_n\tau = (2n + 1)\pi$ $(n = 0, 1, 2 \ldots)$.

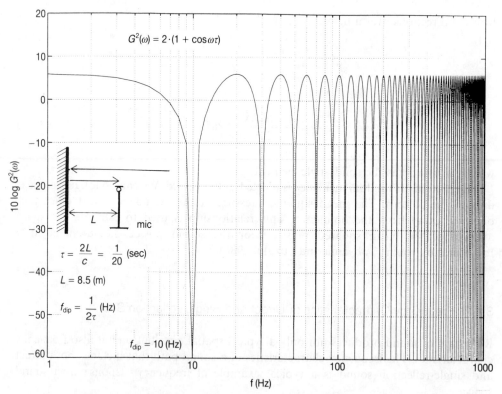

Fig. 3.1.3. Frequency characteristic of the sound path including direct and reflection sounds.

Single reflection delayed more than 30 ms after the direct sound is, in general, perceived separately from the direct sound, like an echo. Spectrum changes caused by reflection delayed less than 30 ms significantly affect the perception. The optimal reflection sounds in a concert hall will be discussed in Chapter 7.

Reflection sounds are frequently harmful, because of the spectrum deformation. Sound recording in a narrow space suffers this problem because of the early reflection from the close walls. The frequency of the zeros becomes higher as the delay time of the reflection becomes shorter. Therefore, if we set the microphone at the wall, the delay time becomes extremely short (almost zero). Consequently the zero frequencies move to the extremely high frequency region out of the audible frequency range. We can therefore eliminate the problems by setting the microphone at the boundary. This type of microphone is available in the market.

Sound paths composed of a finite number of reflections are characterized by 'zeros'. This is the same as the frequency response obtained using a finite impulse response (FIR) digital filter. The impulse response of a sound path composed of direct and single-reflection sound is the simplest example of an FIR filter. The frequency characteristic of a sound path is obtained by Fourier transform from the impulse response as described in Section 2.1.1. As shown in Fig. 3.1.3, the zeros are periodically repeated in a comb pattern. An FIR filter is therefore frequently called a comb filter.

3.2. Sound Power Output of a Source

The sound power of a source is an important factor in sound space control. The most essential element in creating a quiet space is reduction of any machine noise from the sound source. The sound power radiated from a source, however, varies owing to environmental conditions, and differs from the free-field value. It is therefore often desirable to estimate the power output of the sound source in a reflecting environment. Waterhouse (1963) theoretically analysed the effects of reflectors on the sound power output of a source. Also see Sections 4.3.1 and 4.3.2. We will review Waterhouse's approach in Section 3.2.2. The method used to measure the sound power of a source will be described in Chapter 6.

3.2.1. Sound Power Output Calculation

The sound power is the sound energy radiated by a source per unit of time. The sound wave radiated from a pulsating spherical source is a spherical wave. If we assume an $\exp(j\omega t)$ time dependency, the spherical wave equation, the velocity potential, the sound pressure, and the particle velocity at point distant r from a pulsating sphere of radius a that has a volume velocity $Q\exp(j\omega t)$ (m³/s), are respectively,

$$\frac{\partial^2(r\phi)}{\partial r^2} + k^2(r\phi) = 0 \quad \text{(m/s)}$$

$$\phi = \frac{A}{4\pi r}e^{-jkr} \quad \text{(m}^2\text{/s)}$$

$$p = \kappa s = \rho\frac{\partial\phi}{\partial t} \quad \text{(Pa)}$$

and the particle velocity $\dot{\xi}$ is (leaving the time dependent term)

$$\dot{\xi} = \phi \frac{1 + jkr}{r} \quad \text{(m/s)} \tag{3.2.1a}$$

where ρ is the density of air (kg/m^3), k is the wavenumber (1/m), κ is the bulk modulus (Pa), s is the condensation, c is the speed of sound in the air (m/s), $c = \sqrt{\kappa/\rho}$ (m/s), and

$$A = \frac{Q}{1 + jka} e^{jka} \quad \text{(m}^3\text{/s)} \tag{3.2.1b}$$

Taking a surface Σ that has a radius of r from the centre of the source in a free field, the sound power radiated from a pulsating sphere is given by

$$P = \frac{1}{2} \text{Re} \int_{\Sigma} p\dot{\xi}* \, ds = \frac{\rho c Q^2}{8\pi a^2} \frac{k^2 a^2}{1 + k^2 a^2} \quad \text{(W)} \tag{3.2.2}$$

where Re($\cdot$) denotes the real part of ($\cdot$) and * denotes complex conjugate. Thus, the sound power of a monopole (point) source ($ka \to 0$) in a free field is given by

$$P_0 = \frac{\rho \omega^2 Q^2}{8\pi c} \quad \text{(W)} \tag{3.2.3}$$

We define the radiation efficiency of a sphere as (Lyon, 1987)

$$\sigma_{\text{rad}} = \frac{k^2 a^2}{1 + k^2 a^2} \tag{3.2.4}$$

3.2.2. Sound Power Output in a Reflecting Environment

A. Radiation Impedance

Following Waterhouse (1963), we wish to calculate the power output of a point source in a reflecting environment. The general way to do this is to evaluate

$$\int \overline{p\dot{\xi}} \, ds \tag{3.2.5}$$

over any convenient surface surrounding the source, where p and $\dot{\xi}$ are the pressure and particle velocity of the composite sound field composed of the direct and reflected waves and $^-$ shows the time average. This procedure can be simplified by reducing the integral to a summation of pressures at one point, the point occupied by the point source.

The acoustic impedance for the source, is defined as the pressure at the source surface divided by the volume velocity through the surface. This impedance function (Pa s/m^3), is a complex scalar quantity and useful for describing the reaction of a reflecting environment on a sound source. For a pulsating-sphere source in a free field, the acoustic (radiation) impedance is

$$z_{\text{rad}} = \frac{\rho c}{4\pi a^2} \frac{k^2 a^2 + jka}{1 + k^2 a^2} = R_{\text{rad}} + jX_{\text{rad}} \quad \text{(Pa s/m}^3\text{)} \tag{3.2.6}$$

Similarly, for a monopole source ($ka \to$ small) it is

$$z_{0rad} = \frac{\rho c}{4\pi a^2}(k^2 a^2 + jka) = R_{0rad} + jX_{0rad} \quad (\text{Pa s/m}^3).$$ (3.2.7)

The real part of this impedance is proportional to the power radiated. For a pulsating sphere as shown by Eq. (3.2.2)

$$P = \frac{\rho c Q^2}{8\pi a^2}\frac{k^2 a^2}{1 + k^2 a^2} = \tfrac{1}{2}Q^2 \,\text{Re}(z_{rad}) \quad (\text{W})$$ (3.2.8a)

and for a monopole source

$$P_0 = \frac{\rho \omega^2 Q^2}{8\pi c} = \tfrac{1}{2}Q^2 \,\text{Re}(z_{0rad}) \quad (\text{W})$$ (3.2.8b)

as Eq. (3.2.3).

Only the real part of the radiation impedance contributes to the sound power radiation. The imaginary part is frequently called 'an additonal mass' attached to a source. Using the mechanical radiation impedance z_{mrad} (N s/m), the imaginary part of the mechanical radiation impedance is expressed like that for the pulsating sphere,

$$X_{mrad} = 4\pi a^2 \rho c \frac{ka}{1 + k^2 a^2} = \omega \rho \frac{4\pi a^3}{3}\left(\frac{3}{1 + k^2 a^2}\right)$$

$$= \omega M_{add} \quad (\text{N s/m})$$

where

$$M_{add} = \rho \frac{4\pi a^3}{3}\left(\frac{3}{1 + k^2 a^2}\right) \quad (\text{kg})$$ (3.2.9)

When the sound power is radiated from a pulsating sphere source, the source forces the surrounding medium, whose volume is three times that of source itself at low frequencies.

B. *Radiation Impedance in a Reflecting Environment*

When a small, pulsating spherical source is placed at a point O in the reflecting environment, the acoustic radiation impedance is defined as, using Eq. (3.2.7),

$$z_{rad} = z_{0rad} + z_r$$ (3.2.10)

where z_r is the 'reflected impedance', which shows the effect of the reflections from the boundaries on the source. Except for the case where the reflected waves are focused back to the source, the volume velocity of the reflected waves at the surface of the source vanishes as $ka \to 0$. This means that at the source point, the reflected impedance is simply

$$z_r = \frac{\sum p_r}{Q} \quad (\text{Pa s/m}^3)$$ (3.2.11)

which is the ratio of the sum of the pressures of the reflected waves to the volume velocity of the point source. The power output P of the point source in a reflecting environment is therefore

$$P = \tfrac{1}{2} Q^2 (R_{0\text{rad}} + R_r) \quad (\text{W})$$ (3.2.12)

where $R_{0\text{rad}}$ and R_r are the real part of $z_{0\text{rad}}$ and z_r, respectively.

The fractional change in the source output when it is moved from a free field to a point in a reflecting environment is

$$\frac{P}{P_0} = 1 + \frac{R_r}{R_{0\text{rad}}},$$ (3.2.13)

where P_0 is the sound power of the monopole source in a free field given by Eq. (3.2.3). R_r can be either negative or positive and the power output of the source can either decrease or increase in reaction to the environment. Source power emission is limited by the supplied power and the source volume velocity decreases when the source encounters an extremely high impedance. R_r can become infinitely large but it cannot be less than $-R_{0\text{rad}}$, for this would correspond to the source becoming a sink thereby taking power from the environment. R_r can be controlled electrically using secondary loudspeakers. This type of sound field control is called active power minimization. We will discuss it in Sections 4.3.2 and 6.4.4.

3.2.3. Sound Power Output over a Reflecting Plane

Similarly to the single reflection discussed in Section 3.1, the reaction caused by a reflecting plane is also a simple example of the change in sound power radiation in a reflecting environment. Suppose the plane is perfectly rigid, as shown in Fig. 3.2.1. The boundary condition at the plane surface is that the normal fluid velocity is zero. We may replace the effect of the boundary plane by using a set of image sources, symmetrically placed with respect to the boundary plane, with both the source and its image radiating into unbounded space.

A. *Image Source*

If the source is the simple source at point S shown in Fig. 3.2.1, the image source will also be a simple source, of equal strength and phase, at point S'. The acoustic pressure at point O above the xy plane for the source is

$$p(O) = j\omega\rho Q \left(\frac{1}{4\pi r} e^{-jkr} + \frac{1}{4\pi r'} e^{-jkr'} \right) \quad (\text{Pa}).$$ (3.2.14)

This represents, in the region $h > 0$, the original wave plus a reflected wave, which satisfies the boundary conditions at $h = 0$. Along the boundary surface, the pressure is just twice what it would have been if the boundary had not been present.

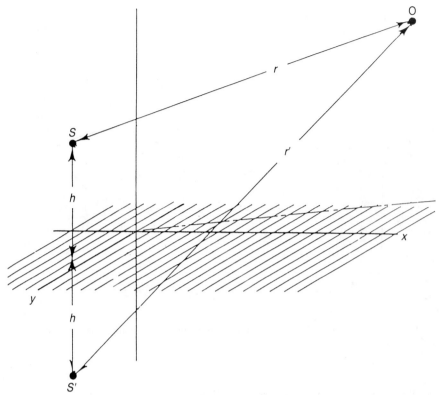

Fig. 3.2.1. Sound source at S, image source at S', and observation point O over a xy-reflecting plane. After Morse and Ingard (1986).

B. *Acoustic Radiation Impedance and Power Output*

The real part of the acoustic radiation impedance for a point source above a reflector is

$$R_{rad} = R_{0rad} + R_r = \frac{\omega k \rho}{4\pi}\left(1 + \frac{\sin(2kh)}{2kh}\right) \quad (\text{Pa s/m}^3) \qquad (3.2.15)$$

The sound power output varies, as the position of the source moves from the reflector. The radiation efficiency is as shown in Eq. (3.2.13),

$$\frac{P}{P_0} = 1 + \frac{R_r}{R_{0rad}} = 1 + \frac{\sin(2kh)}{2kh} \qquad (3.2.16)$$

This is illustrated in Fig. 3.2.2. When $kh = 2\pi h/\lambda$ is small, which occurs either when the source is close to the wall or the wavelength is long, the impedance is twice that for a source in a free field. For long wavelengths, the image source reinforces the primary source. As kh increases, the efficiency approaches unity as shown in Fig. 3.2.2. For a very high-frequency source, the source radiates as though the boundary were not present.

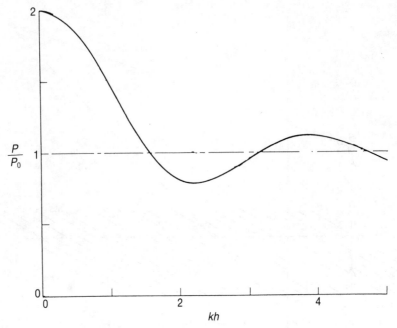

Fig. 3.2.2. Radiation impedance of a point source over a reflecting plane. After Morse and Ingard (1986).

3.2.4. Radiated Sound from a Vibrating Plate

Sound radiation from a vibrating plate and sound transmission through a wall are important issues for noise control in enclosed spaces. The sound transmission loss through a single wall, which is defined as the inverse of the sound energy transmission coefficient, is estimated using the mass law (London, 1945).

A. *Transmission Loss of an Infinitely Extended Single Wall*

Suppose that a plane wave meets an infinitely extended thin wall as shown in Fig. 3.2.3. Assuming that the mechanical impedance per unit area of the wall is z_m(Pa s/m), we have the following equations:

$$(p_i + p_r) - p_t = z_m \dot{\zeta} \quad (\text{Pa}),$$

$$\frac{p_i}{\rho c} \cos \theta - \frac{p_r}{\rho c} \cos \theta = \frac{p_t}{\rho c} \cos \theta = \dot{\zeta} \quad (\text{m/s})$$

$$p_t = \frac{\rho c \dot{\zeta}}{\cos \theta} = z_{\text{mrad}} \dot{\zeta} \quad (\text{Pa}), \tag{3.2.17}$$

and

$$z_{\text{mrad}} = \frac{\rho c}{\cos \theta} \quad (\text{Pa s/m}) \tag{3.2.18}$$

where $\dot{\zeta}$ denotes the normal velocity to the surface of the wall (m/s), z_{mrad} is the mechanical radiation impedance per wall unit area (Pa s/m), θ denotes the incident

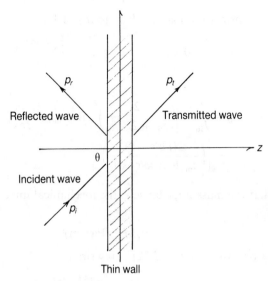

Fig. 3.2.3. Sound wave transmission through a thin wall.

angle (rad) of the plane wave, and p_i, p_r and p_t are the sound pressures (Pa) at the wall of the incident, reflected and transmitted waves, respectively. The boundary conditions assumed here are that (1) wall vibration is determined by the pressure difference between the two sides of the wall, and (2) the normal velocity to the wall surface is continuous at the wall.

The normal velocity of the wall excited by an incidence plane wave is given by both the mechanical and radiation impedance quantities,

$$\dot{\zeta} = \frac{p_i + p_r}{z_m + z_{mrad}} = \frac{2p_i - p_t}{z_m + z_{mrad}} = \frac{2p_i}{z_m + 2z_{mrad}} \quad \text{(m/s)} \tag{3.2.19}$$

Equation (3.2.18) shows that the vibration velocity of the wall is determined by the incident sound pressure (Pa), the unit area mechanical impedance of the wall (Pa s/m), and the unit area mechanical radiation impedance (Pa s/m).

From Eq. (3.2.19) we have the sound power radiated from a unit area of the wall, J (W/m^2):

$$J = \tfrac{1}{2}|\dot{\zeta}^2| \operatorname{Re}(z_{mrad})$$

$$= \frac{1}{2}|p_i^2| \left| \left(\frac{2}{z_m + 2z_{mrad}} \right)^2 \right| \operatorname{Re}(z_{mrad}) \quad \text{(W/m}^2\text{)} \tag{3.2.20}$$

The incident power to the unit area of the wall is given by

$$J_{in} = \frac{1}{2} \frac{|p_i^2|}{\rho c} \cos \theta \quad \text{(W/m}^2\text{)} \tag{3.2.21}$$

Consequently, the transmission loss in dB in terms of $1/\tau_\theta$,

$$\mathrm{TL}_\theta = 10\log\frac{1}{\tau_\theta} = 10\log\left|\left(1 + \frac{z_\mathrm{m}\cos\theta}{2\rho c}\right)^2\right| \tag{3.2.22}$$

where

$$\tau_\theta = \frac{J}{J_\mathrm{in}} = \left|\left(\frac{2}{z_\mathrm{m} + 2z_\mathrm{mrad}}\right)^2\right| \mathrm{Re}(z_\mathrm{mrad})\frac{\rho c}{\cos\theta}$$

$$= \left|\left(\frac{2(\rho c/\cos\theta)}{z_\mathrm{m} + 2(\rho c/\cos\theta)}\right)^2\right| \tag{3.2.23}$$

If we assume that the mass impedance is the mechanical impedance per unit area of the wall,

$$z_\mathrm{m} = j\omega M \quad (\mathrm{Pa\,s/m}) \tag{3.2.24}$$

then the transmission loss by Eq. (3.2.22) is rewritten as

$$\mathrm{TL}_\theta = 10\log\left|\left(1 + \frac{j\omega M\cos\theta}{2\rho c}\right)^2\right| \tag{3.2.25}$$

where M is the mass of the unit area of the wall ($\mathrm{kg/m^2}$). We call Eq. (3.2.25) the 'mass law' equation because the transmission loss is estimated using only the mass density and the frequency. The transmission loss decreases as the incidence angle of the plane wave increases. In the case of grazing (tangential incidence), the transmission loss is 0 dB at the limit.

This theoretical estimation based on the mass law can be extended to the random incidence condition (London, 1945). The transmission loss of a single wall follows the mass law for the most part; however, the transmission loss at high frequencies is often considerably less than that by the mass law. This is called the coincidence effect, which is again related to the sound power radiation efficiency from vibrating structures (Cremer, 1949). We will describe the transmission loss of a vibrating structure, such as a plate, in the next section.

B. *Sound Transmission through a Bending Plate*

(1) *Speed of the Bending Wave (Lyon, 1987).* The free motion of a bending plate is expressed by

$$B\frac{\partial^4\zeta}{\partial x^4} = -M\frac{\partial^2\zeta}{\partial t^2} \quad (\mathrm{Pa}), \tag{3.2.26}$$

where ζ denotes the displacement (m) at time t at point x on the plate, M is the mass per unit area ($\mathrm{kg/m^2}$) of the plate, and B is the bending rigidity ($\mathrm{N\,m}$) written as

$$B = \frac{E}{1-\sigma^2}\frac{h^3}{12} \tag{3.2.27}$$

where

E: Young's modulus ($\mathrm{N/m^2}$)

h: thickness of the plate (m)

σ: Poisson's ratio

We assume that our solution is in the form exp($j\omega t$) and introduce as the bending wavenumber $k_b = \sqrt{\omega}(M/B)^{1/4}$ (1/m). The wave propagation speed is

$$c_b = \frac{\omega}{k_b} = \sqrt{\omega}\left(\frac{B}{M}\right)^{1/4} \quad \text{(m/s)} \tag{3.2.28}$$

The speed of a bending wave depends on its frequency, while all the compressional waves travel in the air at the same speed. Since we can think of any waveform as a summation of its frequency components, we can see that the different frequency components of a wave will travel at different speeds and therefore the shape of the wave will distort and change as the wave moves. This phenomenon is known as dispersion.

(2) *Sound Transmission through an Infinitely Extended Bending Plate.* Suppose again that a plane wave meets an infinitely extended thin plate as shown in Fig. 3.2.4. The incidence angle of the coming plane wave is θ (rad) and its wavelength is λ (m). The excited wave motion on the bending plate has a wavelength of $\lambda_1 = \lambda/\sin\theta$. The displacement of the bending wave is expressed as

$$\zeta = \zeta_0 e^{j\omega t - jk_1 x} \quad \text{(m)} \tag{3.2.29}$$

where k_1 is the wavenumber of the excited wave motion on the plate, and is given by

$$k_1 = \frac{\omega}{c_1} = \frac{2\pi}{\lambda_1} = \frac{\omega}{c}\sin\theta \quad \text{(1/m)} \tag{3.2.30}$$

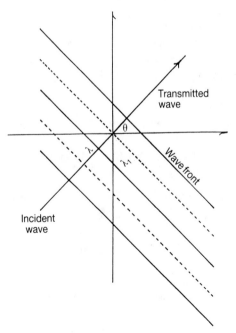

Fig. 3.2.4. Sound transmission through a thin bending plate. $\lambda =$ wavelength of an incident wave; $\lambda_1 =$ wavelength of the excited wave on the plate; $\theta =$ incident angle.

The motion of the plate is expressed by

$$B \frac{\partial^4 \zeta}{\partial x^4} + M \frac{\partial^2 \zeta}{\partial t^2} = (p_i + p_r) - p_t \quad \text{(Pa)} \tag{3.2.31}$$

and

$$\frac{p_i}{\rho c} \cos \theta - \frac{p_r}{\rho c} \cos \theta = \frac{p_t}{\rho c} \cos \theta = \dot{\zeta} \quad \text{(m/s)}$$

Consequently, we have

$$B \frac{\partial^4 \zeta}{\partial x^4} + M \frac{\partial^2 \zeta}{\partial t^2} = 2p_i - 2p_t \quad \text{(Pa)} \tag{3.2.32}$$

Here we assume the plate is sufficiently thin compared to the wavelength of the waves excited by the plate motion.

Following these equations, we have

$$j\omega M \left(1 - \frac{B\omega^2 \sin^4 \theta}{Mc^4}\right) \zeta = j\omega M \left(1 - \frac{B\omega^2}{Mc_1^4}\right) \zeta$$

$$= j\omega M \left(1 - \frac{c_b^4}{c_1^4}\right) \zeta = 2(p_i - p_t) \quad \text{(Pa)}, \tag{3.2.33}$$

c_b denotes the speed of the bending wave (m/s) given by Eq. (3.2.28). Thus, the velocity of the bending wave is

$$\dot{\zeta} = \frac{2(p_i - p_t)}{j\omega M \left(1 - \frac{c_b^4}{c_1^4}\right)} = \frac{2p_i}{j\omega M \left(1 - \frac{c_b^4}{c_1^4}\right) + 2z_{\text{mrad}}} \quad \text{(m/s)} \tag{3.2.34}$$

and

$$z_m = j\omega M \left(1 - \frac{c_b^4}{c_1^4}\right) \quad \text{(Pa s/m)} \tag{3.2.35}$$

We can determine the frequency at which $c_b = c_1$ using Eq. (3.2.28).

$$f_{\min} = \frac{c^2}{2\pi} \sqrt{\frac{M}{B} \frac{1}{\sin^2 \theta}} \quad \text{(Hz)} \tag{3.2.36}$$

When the frequency of the incident plane wave is $f_{\min}$, the plate velocity magnitude is at its maximum, and the transmission loss of the plate is

$$TL_\theta = 10 \log \left| \left[1 + j\omega M \left(1 - \frac{c_b^4}{c_1^4}\right) \frac{\cos \theta}{2\rho c} \right]^2 \right| \quad \text{(dB)} \tag{3.2.37}$$

TL becomes $0\,(\mathrm{dB})$ at $f_{\min}\,(\mathrm{Hz})$. The lowest $f_{\min}$ frequency is normally called the coincidence frequency and is defined at an incident angle of $\pi/2$,

$$f_c = \frac{c^2}{2\pi}\sqrt{\frac{M}{B}} = \frac{c^2}{2\pi}\sqrt{\frac{M}{Eh^3/12(1-\sigma^2)}} \quad (\mathrm{Hz}) \tag{3.2.38}$$

The transmission loss of the plate decreases at the coincidence frequency, even if the size of the plate is finite. The coincidence effect on the transmission loss of a finite plate was investigated by Tohyama and Itow (1974).

(3) *Sound Radiation from an Infinite Plate Bending Wave.* The radiation impedance of sound transmitted through a wall always has only a real part, as described above. Thus, the radiated power is proportional to the amplitude of the vibration of the plate when the angle of incidence of the incoming wave is fixed. We now consider the vibration of a thin plate excited by a mechanical force such as a line force. The radiation impedance in this case depends on the frequency, and is a complex variable.

Suppose that a travelling bending wave is excited on a plate as shown in Fig. 3.2.5. The amplitude of its vibration velocity is expressed as

$$\dot{\zeta}_0 = A e^{-k_b x} \quad (\mathrm{m/s}) \tag{3.2.39}$$

Similarly to Eq. (3.2.18), the mechanical radiation impedance of a unit area becomes

$$z_{\mathrm{mrad}} = \frac{k\rho c}{\sqrt{k^2 - k_b^2}} = \frac{\lambda_b \rho c}{\sqrt{\lambda_b^2 - \lambda^2}} \quad (\mathrm{Pa\,s/m}) \tag{3.2.40}$$

and

$$\lambda_b = \frac{c_b}{f} \qquad \lambda = \frac{c}{f} \quad (\mathrm{m}) \tag{3.2.41}$$

where $\lambda_b\,(\mathrm{m})$ is the wavelength of the bending wave excited on the plate. When the speed of the bending wave is faster than the speed of sound in the air, the radiation impedance is real. On the other hand the radiation impedance becomes imaginary when the speed of the bending wave is slower than the speed of sound. The radiation impedance becomes infinite as the speed of both waves becomes equal at the coincidence frequency. If the radiation impedance is imaginary, sound energy is not radiated from the vibrating plate into the surrounding air.

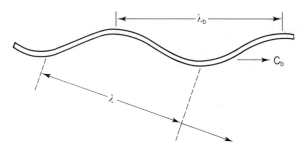

Fig. 3.2.5. Bending wave and sound radiation. From Lyon (1987).

3.3. Room Reverberation Theory

Reverberation control is the most fundamental issue in sound field technology. Current room acoustic technology is still focused on estimating reverberation time. In this chapter we will review the classical reverberation theories and also discuss some modern investigations of the reverberation process. Classical room reverberation theory was primarily developed by Sabine (1922) and Eyring (1930). Those theories are based on the randomness of sound fields resulting from complex wave phenomena in an enclosed space.

3.3.1. Random Process and Reverberation Theories

The sound path from a source to a listener is, in principle, deterministically predictable in an enclosed space. The time history of reverberant sound reflection is expressed by the impulse response between a source and a receiving point. An impulse response in a room is composed of the direct sound, the early reflection sounds, and the reverberation sounds. These three components affect the auditory senses of the listener quite differently, as is shown in other sections of this book. Reverberation theory, usually describes the energy decay process of the reverberation sound.

The impulse response is highly sensitive to the location of both the source and the receiver. It may be appropriate to assume that samples of impulse responses at different locations are an ensemble of stochastic processes. Schroeder (1965a) and Kuttruff (1958) introduced probability distributions of the number of reflections. Classical reverberation theory is expressed using the probability distributions, which are defined by the mean free path or the collision frequency of the reverberation sound into the surroundings.

A. Binomial Distribution of Reflections and Eyring's Formula

Suppose that the boundaries of the room are composed of K different surfaces, where α_k is the absorption coefficient of the kth surface, and n_k is the number of sound ray reflections from the kth surface in the time interval t. The reverberation process of the reverberant sound energy density can then be written as (Kuttruff, 1958)

$$E(t) = E_0 \prod_{k=1}^{K} (1 - \alpha_k)^{n_k} \quad (\text{J/m}^3) \tag{3.3.1}$$

where E_0 is the energy density in the room at a steady state ($t = 0$, i.e. the time when the sound source stops). We introduce here a probability distribution for the n_k. The ensemble averaged reverberation process is expressed by

$$E(t) = E_0 \sum_{n_1, n_2, \ldots} p(n_1, n_2, \ldots, n_k) \prod_{k=1}^{K} (1 - \alpha_k)^{n_k} \quad (\text{J/m}^3) \tag{3.3.2}$$

where $p(n_1, n_2, \ldots, n_k)$ is the joint probability distribution for the n_k.

The reverberation process depends not only on the absorption conditions at the boundaries but also on the probability distributions of the number of reflections in a particular time interval. Therefore, as Schroeder and Hackman (1980) pointed out, many reverberation formulae can be deduced by introducing different probability distributions for the number of reflections. Here we introduce popular conventional formulae.

(1) *Diffuse Field Assumptions and the Probabilities for the Sound Reflections.* To simplify the reverberation expression, we follow Kuttruff (1958) and assume that the boundary is composed of two parts. One part has an area of S_1 with a sound absorbing coefficient α_1, and the other part has an area of S_2 with a coefficient of α_2. In a diffuse field, we can assume that the sound energy flows equally in all directions. The probability of hitting one portion of the wall is determined by the ratio of the area for that portion of the wall to the total wall area.

Thus, if we assume that subsequent wall reflections are stochastically independent of each other, then the probability of n collisions with wall position S_1 from a total number of reflections N is given by the binomial distribution

$$p_N(n) = {}_NC_n \left(\frac{S_1}{S}\right)^n \left(\frac{S_2}{S}\right)^{N-n} \tag{3.3.3}$$

(2) *Deterministic Total Number of Reflections and Eyring's Reverberation Formula.* After n collisions with S_1 and $N - n$ collisions with S_2, the remaining energy becomes

$$E_N(n) = E_0(1 - \alpha_1)^n (1 - \alpha_2)^{N-n} \quad (\text{J/m}^3) \tag{3.3.4}$$

Using the binomial distribution in Eq. (3.3.3), the expectation of the reverberation energy is expressed by

$$\langle E_N(n) \rangle_n = \sum_{n=0}^{N} p_N(n) E_N(n)$$

$$= E_0 \left(\frac{S_1}{S}(1 - \alpha_1) + \frac{S_2}{S}(1 - \alpha_2)\right)^N \quad (\text{J/m}^3) \tag{3.3.5}$$

Since $S_1 + S_2 = S$, this equation can be written as

$$\langle E_N \rangle_n = E_0(1 - \tilde{\alpha})^N$$

$$= E_0 \exp[N \ln(1 - \tilde{\alpha})] \quad (\text{J/m}^3) \tag{3.3.6}$$

where the average absorption coefficient is

$$\tilde{\alpha} = \frac{1}{S}(S_1\alpha_1 + S_2\alpha_2) \tag{3.3.7}$$

If we assume that the total number of reflections N from all the surfaces, using the mean free path $m = 4V/S$ (Kosten, 1960), is given deterministically as

$$N = \frac{ct}{m} = \frac{cSt}{4V} \tag{3.3.8}$$

then we have Eyring's (1930) formula (Schroeder, 1965a),

$$\langle E_N(n)\rangle_n = E(t) = E_0 \exp\left(\frac{cS}{4V} t \ln(1 - \tilde{\alpha})\right) \quad (\text{J/m}^3) \tag{3.3.9}$$

where V denotes the room volume (m^3) and c is the sound speed (m/s).

Classical reverberation time is defined as the time it takes for the sound energy to decrease 60 dB from that at steady state. Reverberation time is therefore given by

$$T_R = \frac{4V}{c(-\ln(1 - \tilde{\alpha}))S} (\ln 10^6) \approx 0.161 \frac{V}{-\ln(1 - \tilde{\alpha})S} \quad (\text{s}) \tag{3.3.10}$$

B. *Poisson Distribution and Sabine's Formula*

Another historically important formula is Sabine's formula. As we mentioned above, a different probability distribution produces a different reverberation formula. In the binomial distribution above, N is exactly determined. It may be more appropriate to assume that N is also distributed about some mean value (Schroeder, 1965a).

There is a mathematical family of distribution functions for binomial distributions (Ochi, 1990). One of these distribution functions is the Poisson distribution. The following Poisson distribution is obtained as the limit of a binomial distribution as N becomes large but the expectation $\tilde{n}$ remains finite:

$$p(n) = \left(\frac{\tilde{n}^n}{n!}\right) e^{-\tilde{n}} \tag{3.3.11}$$

If we assume independent Poisson distributions for the n_k with the average

$$\tilde{n}_k = \frac{cS_k}{4V} t, \qquad p(n_1, n_2, \ldots, n_k) = \prod_{k=1}^{K} \left(\frac{\tilde{n}_k^{n_k}}{n_k!}\right) e^{-\tilde{n}_k} \tag{3.3.12}$$

Sabine's formula becomes

$$E(t) = E_0 \sum_{n_1, n_2, \ldots} p(n_1, n_2, \ldots, n_k) \prod_{k=1}^{K} (1 - \alpha_k)^{n_k}$$

$$= E_0 \exp\left(-\frac{c}{4V} t \sum_k \alpha_k S_k\right) \quad (\text{J/m}^3) \tag{3.3.13}$$

Consequently, reverberation time is

$$T_R \approx 0.161 \frac{V}{\bar{\alpha}S} \quad (\text{s}) \tag{3.3.14}$$

C. *Comparison of Sabine's and Eyring's Formulae*

Both Sabine's and Eyring's formulae are still frequently used for acoustic design because they can be simply expressed using only geometric parameters. Sabine's formula, however, has a problem in the limit case.

As the sound absorption coefficient approaches unity, the reverberation sound energy and the reverberation time estimated from Eqs (3.3.1) and (3.3.2) become zero. Sabine's reverberation formula still gives a finite reverberation time in the limit case of Eq. (3.3.14). With Eyring's formula, however, the reverberation time decreases continuously to zero as the absorption coefficient approaches unity. Eyring's formula is preferable to Sabine's formula.

D. *Wave Theory and Power-Law Decay Formula*

Schroeder (1965a) introduced a decay formula based on statistical wave theory. In the wave theory, many wave modes (see Sections 3.4 and 3.5) are in a frequency range. The ensemble averaged energy decay function for a frequency interval is therefore given by an integration of exponential functions with different decay rates:

$$E(t) = E_0 \int_0^\infty p(\gamma) e^{-\gamma t} d\gamma \quad (J/m^3) \tag{3.3.15}$$

where γ denotes a decay constant and $p(\gamma)$ is the distribution function for the decay rate. Suppose that there is only one type of absorber on the room surfaces and that the absorption on the uncovered surfaces and in the air is negligible. The amount of absorbed energy at the boundaries is proportional to the incident energy at the boundary area. If we take an area S_{ab} covered by an absorbing material, the total incident energy flux, J, can be expressed as

$$J \sim \sum_{n=1}^{N_s} p_n^2 \tag{3.3.16}$$

and the decay constant can be written as

$$\gamma \sim \sum_{n=1}^{N_s} p_n^2 \tag{3.3.17}$$

where p_n^2 denotes the squared sound pressure data sampled at a point in area S_{ab}, N_s shows the total number of independent samples in the data, $N_s \approx S/(\lambda/2)^2$ (see Section 3.4), and λ is the wavelength of the sound wave of interest.

As shown in Section 3.4, the square pressure data sampled in a diffuse field are distributed according to an exponential distribution. As shown by Eq. (3.3.17), the decay constant written as a summation of N_s independent exponentially distributed variables therefore follows an N_sth order Gamma-distribution,

$$p(\gamma) = N_s \frac{(N_s \gamma/\tilde{\gamma})^{N_s-1}}{(N_s - 1)! \tilde{\gamma}} e^{-N_s \gamma/\tilde{\gamma}} \tag{3.3.18}$$

where $\tilde{\gamma}$ denotes the average decay constant. By introducing this Gamma-distribution function into Eq. (3.3.15), we obtain a power law for the decay formula:

$$E(t) = E_0 \left(1 + \frac{\tilde{\gamma} t}{N_s}\right)^{-N_s} \quad (J/m^3) \tag{3.3.19}$$

This formula can be written asymptotically as a simple exponential decay function:

$$E(t) \approx E_0 e^{-\tilde{\gamma} t} \quad (J/m^3) \tag{3.3.20}$$

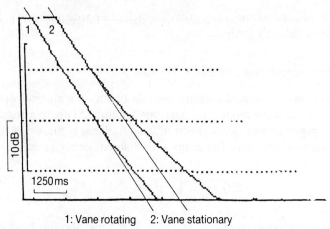

1: Vane rotating   2: Vane stationary

Fig. 3.3.1. Sample of ensemble averaged decay curves. Centre frequency 100 Hz. From Kawakami and Yamaguchi (1986).

as N_s becomes large so that $\tilde{\gamma}t/N_s \ll 1$. Measuring the level in decibels, the energy level decays as

$$L(t) = -4.34 N_s \ln\left(1 - \frac{\tilde{\gamma}t}{N_s}\right)$$

$$\approx -4.34\tilde{\gamma}t \quad \text{dB} \quad (\tilde{\gamma}t/N_s \ll 1) \tag{3.3.21}$$

This power law formula suggests that energy decay generally does not follow either Eyring's or Sabine's conventional exponential formulae. We can expect the conventional decay process when the frequency of interest becomes high enough that we can take a large number of independent N_s samples on the absorbing material surface. The decay rate, defined as $\dot{E}(t)/E(t)$, is obtained from Eq. 3.3.19,

$$-\left(\frac{\dot{E}(t)}{E(t)}\right)^{-1} = \frac{1}{\tilde{\gamma}} + \frac{t}{N_s}$$

$$\rightarrow \frac{1}{\tilde{\gamma}} \quad (t \rightarrow 0, \text{ or } N_s \rightarrow \text{large}) \tag{3.3.22}$$

This decay rate formula shows that the initial decay property gives the absorption coefficient of the material, if the decay formula does not follow the conventional exponential function (Tohyama, 1986). Kawakami and Yamaguchi (1986) have developed an effective method for measuring the sound absorption coefficients of materials following the power-law decay formula (Fig. 3.3.1).

3.3.2. Two-Dimensional Diffuse Fields and Nonexponential Reverberation Decay

Suppose that the floor and ceiling of a room are covered with absorbing material, while the side walls, perpendicular to the floor, are acoustically hard. This is typical for concert halls. In such rooms, the reverberation time cannot be estimated using classical reverberation theory because that assumes equal probabilities for sound incidence at any part of the boundary for any time interval. That is, the effects of the locations of the sound-absorbing materials on the reverberation process are not taken into account in the classical theories. In reverberation theory, therefore, the effects of the absorbing materials and their arrangement and of the room shape on the reverberation process have been more thoroughly investigated.

We describe in this section a hyperbolic type of decay function in a 2-dimensional field based on image theory (Barron, 1973; Houtgast, private communication). The reverberation process under this condition is not analysed well by conventional theories (Kosten, 1960). A non-exponential decay formula has already been introduced by Schroeder (1965a). It assumes a superposition of exponential functions with different decay rates. The hyperbolic formula that we introduce here, however, is not derived from the superposition formula of exponential functions by changing only the distribution law of sound reflections.

A. *Two-dimensional Array of Image Sources*

The physical space inside boundaries is normally 3-dimensional. In acoustics however, we also have the cases of 2- (or even 1-) dimensional fields. A 2-dimensional sound field, for example, occurs in a tall building, a highway tunnel, and in a concert hall with hard walls perpendicular to an absorbent floor. Most of their image sources can be arranged in a 2-dimensional plane including the sound source. Figure 3.3.2 shows a possible arrangement of the virtual sources. Suppose that an impulse sound wave is radiated from a source at $t = 0$. The impulse response is determined by the locations of the image sources. Each image source radiates an impulsive spherical wave that has an appropriate time delay and amplitude. The average density of the image sources that radiate the sound energy arriving at a receiving point after the direct sound is

$$n(t) = \frac{2\pi c^2 t}{S} \quad (1/\text{s}) \tag{3.3.23}$$

where S is the area of the room (m^2). The average number of reflections, N, is

$$N = \frac{ct}{m} = \frac{Lct}{\pi S} \tag{3.3.24a}$$

at t where m denotes the mean free path in a 2-dimensional diffuse field (Kosten, 1960),

$$m = \frac{\pi S}{L} \quad (\text{m}) \tag{3.3.24b}$$

L (m) is the total surrounding length of the 2-dimensional space.

B. *Reverberant Energy Decay Formula*

The reverberation decay is the process after the sound source stops. The intensity of a point source is given by

$$J_0 = \frac{P_0}{4\pi r^2} \quad (\text{W/m}^2) \tag{3.3.25}$$

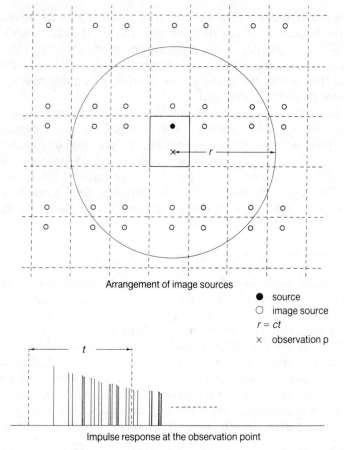

Arrangement of image sources

● source
○ image source
$r = ct$
× observation p

Impulse response at the observation point

Fig. 3.3.2. Arrangement of image sources and impulse response at the observation point.

at a distance r (m) from the point source with power output P_0 (W). The intensity decreases in inverse-square proportion to the distance between the receiving point and the image sources, and it also decreases as the sound power of the source decreases.

The intensity of the reverberation sounds in a time interval of dt is expressed as

$$J(t) = \frac{P_0(1 - \tilde{\alpha})^N}{4\pi(ct)^2} \, n(t) \, dt$$

$$= \frac{P_0(1 - \tilde{\alpha})^N}{2St} \, dt \quad (\text{W/m}^2) \tag{3.3.26}$$

where $\tilde{\alpha}$ denotes the averaged absorption coefficient. Consequently, the reverberant sound energy decay curve becomes (Schroeder 1965b), using Eq. (3.3.24a),

$$E(t) = \frac{P_0}{2cS} \int_t^\infty \exp\left(\ln(1 - \tilde{\alpha}) \frac{Lct}{\pi S}\right) \frac{dt}{t} \quad (\text{J/m}^3). \tag{3.3.27}$$

This decay formula is not a simple exponential function, but is a hyperbolic type.

C. *Initial Decay in a Two-dimensional Reverberation Field*

The energy decay formula Eq. (3.3.27) can be written asymptotically as

$$E(t) \sim \int_t^\infty \frac{dt}{t} \qquad (t \to 0)$$

$$\sim \int_t^\infty \exp\left(\ln(1 - \tilde{\alpha})\frac{Lct}{\pi S}\right) dt \quad (\text{J/m}^3) \qquad (t \to \text{large}) \tag{3.3.28}$$

Thus, the decay rate becomes constant and independent of time t as t becomes large. The reverberation time determined by 'the later part of the energy decay' therefore becomes

$$T_R = \frac{0.128 S}{-\ln(1 - \tilde{\alpha})L} \quad (\text{s}) \tag{3.3.29}$$

This is called the reverberation time in a 2-dimensional diffuse field.

The initial decay rate in a 2-dimensional field, however, becomes very steep, since it depends on t as a hyperbolic function $1/t$. This hyperbolic decay rate independent of the sound absorption is a typical characteristic of a 2-dimensional reverberant space. As mentioned in Section 3.3.1, Kawakami and Yamaguchi (1986) have developed a method for estimating the sound absorption coefficients in a reverberant space from the initial decay rate based on Schroeder's (1965a) theory. The hyperbolic initial decay Eq. (3.3.28) implies, however, it is not appropriate to fully cover the floor by absorbing materials. Such an arrangement of absorption materials results in a hyperbolic type of initial slope independent of the absorption conditions. The initial decay rate is also important for speech intelligibility in a reverberant space, as will be described in Section 7.1.3. The steep initial slope suggests a possibility for improving speech intelligibility without shortening the reverberation time.

3.3.3. Reverberation Time in an Almost-two-dimensional Reverberation Field

We expect, even in a small regular-shaped room with a carpet-covered floor, that the 2-dimensional reverberation formula can be used. As Hirata (1979, 1982b) pointed out, however, in the case of a small room we have to assume that the sound field is 'between' a 2-dimensional and a 3-dimensional field (we call this an 'almost-two-dimensional' field), because the side walls are not very high. Under this type of condition, the feature of reverberation is typically seen in the mid-frequency range since the 3-dimensional field is dominant at lower frequencies, while the 2-dimensional field is significant at higher frequencies (Hirata, 1979, 1982b; Tohyama and Suzuki, 1987).

A. *Average Number of Reflections per Unit Time*

Batchelder (1964) described the average number of reflections a sound wave undergoes at the walls in a rectangular room. The average number v of reflections of an oblique wave in a rectangular room is given by (Morse and Bolt, 1944)

$$v = c(L_y L_z |\alpha| + L_z L_x |\beta| + L_x L_y |\gamma|)/V$$

$$= v_x + v_y + v_z \quad (1/\text{s}) \tag{3.3.30}$$

where L_x, L_y and L_z are the lengths of the sides of the rectangular room (m), V is the volume of the room (m³), c is the speed of sound in air (m/s), α, β, γ are direction cosines,

$$\alpha = \frac{k_x}{k_0}, \qquad \beta = \frac{k_y}{k_0}, \qquad \gamma = \frac{k_z}{k_0}, \tag{3.3.31}$$

k_0 is the wavenumber of the oblique wave, f_0 is the frequency of the oblique wave,

$$k_0^2 = k_x^2 + k_y^2 + k_z^2 \tag{3.3.32}$$

and v_x, v_y, v_z are the respective average numbers of reflections the sound wave undergoes at the x-, y- and z-walls, respectively. The z-walls are located at $z = 0$ and $z = Lz$ in rectangular coordinates, where the origin is located at one corner; the x- and y-side walls are also located in a similar manner.

B. Number of Reflections and the Wavenumber

The average number of reflections that an oblique wave undergoes at the z-walls, v_z, is given by

$$v_z = \frac{c}{L_z} \frac{k_z}{k_0} \quad (1/s) \tag{3.3.33}$$

Thus, when $v_z = r$ the z-component of the wavenumber, k_z, becomes k_{zr}, that is,

$$k_{zr} = k_0 r L_z / c \quad (1/m) \tag{3.3.34}$$

Setting

$$k_{zr} = \frac{n_r \pi}{L_z} \quad (1/m) \tag{3.3.35}$$

the equation

$$n_r = 2 \left(\frac{L_z}{c}\right)^2 f_0 r \tag{3.3.36}$$

is obtained. When $k_z = 0$, the wave is called a 2-dimensional wave. Here we call a wave "almost-2-dimensional wave", when $k_z \leqslant k_{zr}$. The average number of reflections at the z-walls, v_{zr}, in an oblique wave field, where the z components of the oblique wave-numbers range from 0 to k_{zr}, is therefore given by

$$v_{zr} = r/2 \quad (1/s) \tag{3.3.37}$$

C. Ratio of the Average Number of Reflections at the z-Walls to that at the Other Side Walls

The number of reflections at the z-walls is neglected in a 2-dimensional (xy) field. For small-sized rooms, we derive here an 'almost-two-dimensional diffuse field theory' in which reflections at the z-walls are also taken into account. This field is assumed to be composed of tangential waves, oblique waves, and 'almost-tangential waves'. The average number of reflections at the z-walls, v_{zr}, is given by Eq. (3.3.37). Introducing the mean free path m into the 2-dimensional diffuse field (Kosten, 1960), the average number of reflections at the x- and y-walls in the almost-two-dimensional field, v_{xy}, becomes

$$v_{xy} = \frac{n_r c}{m} \quad (1/s) \tag{3.3.38}$$

since there are n_r groups of almost-xy-tangential waves and the average number of reflections in each wave group is c/m. Consequently, the ratio of the average number of reflections at the z-walls to that at the other side walls μ, becomes

$$\mu = \frac{r/2}{n_r c/m} = \frac{mc}{L_z^2 4 f_0} \tag{3.3.39}$$

This ratio has the frequency as a parameter.

D. *Reverberation Time in the Almost-two-dimensional Reverbation (Diffuse) Field*

We define the reverberation time in a 2-dimensional field as

$$T_R = \frac{0.128S}{-\ln(1 - \widetilde{\alpha_{xy}})L} \quad (s) \tag{3.3.40}$$

where $\widetilde{\alpha_{xy}}$ denotes the average sound absorption coefficient in the xy-two-dimensional field. Using this definition, we can obtain the reverberation time for an 'almost-two-dimensional field'. We do this by replacing $\widetilde{\alpha_{xy}}$ with the averaged absorption coefficient $\widetilde{\alpha_{Al\text{-}xy}}$ in the almost two-dimensional reverberation (diffuse) field. The absorption coefficient $\widetilde{\alpha_{Al\text{-}xy}}$ can be found by using the ratio, μ, of the number of reflections given by Eq. (3.3.39).

The reverberation time in the almost-two-dimensional reverberation field is thus given by

$$T_{Al\text{-}xy} = \frac{0.128S}{-\ln(1 - \widetilde{\alpha_{Al\text{-}xy}})L} \quad (s) \tag{3.3.41}$$

where the averaged absorption coefficient, $\widetilde{\alpha_{Al\text{-}xy}}$ is given by

$$\widetilde{\alpha_{Al\text{-}xy}} = \widetilde{\alpha_{xy}}\left(1 - \frac{v_z}{v_{xy} + v_z}\right) + \widetilde{\alpha_z}\left(\frac{v_z}{v_{xy} + v_z}\right)$$

$$\approx \widetilde{\alpha_{xy}}(1 - \mu) + \widetilde{\alpha_z}\mu \tag{3.3.42}$$

Here $\widetilde{\alpha_z}$ is the averaged absorption coefficient of the z-wall. The reverberation time formula above has frequency as a parameter. This frequency dependency is not due to the frequency characteristics of the absorption materials used in a room. Even if there is no frequency dependency of the absorption materials, frequency characteristics of the reverberation time may still exist. As the frequency increases, μ decreases. For high-frequency bands, the height of the side walls surrounding the 2-dimensional diffuse field is acoustically greater, so the absorption of the z-walls is not significant. For low-frequency bands, however, the absorption of the z-walls becomes significant.

3.3.4. Examples and Measurements of Reverberation Time

A. *Reverberation Time Measurements*

We are interested in a frequency-averaged decay curve since the reverberation process of a pure tone is too sensitive to the measurement conditions. If the excitation signal is a bandpass-filtered noise, for the same enclosure and same observation points, different decay curves are obtained owing to the randomness of the excitation signal. To minimize the effect of these fluctuations, an ensemble average is taken by repeating the measurement many times. The ensemble-averaged decay curve, however, can also be obtained using a single-impulse response measurement (Schroeder, 1965b). If we let $n(t)$ be 'stationary white noise', the autocovariance function (see Section 2.2.3) is

$$\langle n(t_1) \cdot n(t_2) \rangle = N \cdot \delta(t_1 - t_2) \tag{3.3.43}$$

where the brackets denote ensemble average, N is the noise power in unit bandwidth, and $\delta(t_1 - t_2)$ is the Dirac delta function.

Suppose that the bandpass filtered noise is radiated into the enclosure and, when a steady state is reached, the noise at the input to the filter is switched off (at $t = 0$). The signal received at a receiving point is then

$$s(t) = p_0 \int_{-\infty}^{0} n(\eta) h(t - \eta) \, d\eta \quad \text{(Pa)} \tag{3.3.44}$$

where $h(t)$ is the combined impulse response of the system, which consists of the noise filter, amplifiers, transducers and the enclosure between the source and receiving points, and p_0 is the constant for the sound pressure response. Averaging the squared received signal over the ensemble of noise signals and applying Eq. (3.3.43), we obtain

$$\langle s^2(t) \rangle = p_0 \int_{-\infty}^{0} d\xi \int_{-\infty}^{0} \langle n(\xi) n(\eta) \rangle h(t - \xi) h(t - \eta) \, d\eta$$

$$= p_0^2 N \int_{t}^{\infty} h^2(\tau) \, d\tau \quad \text{(Pa}^2\text{)} \tag{3.3.45}$$

Equation (3.3.45) shows that the ensemble average of the decay curves is obtained from only a single measurement of the impulse response.

The impulse response will still be difficult to obtain in practice when one has only a low signal-to-noise ratio (Schroeder, 1985; Ando, 1985; Borish, 1985). Hirata (1982a) developed a method for improving the signal-to-noise ratio in reverberation decay measurement. If we have two impulse response data, h_1 and h_2, which are measured at the same position, the signal-to-noise ratio is greatly improved. We multiply the impulse response data pair,

$$\int_{t}^{\infty} h_1(\tau) h_2(\tau) \, d\tau = \int_{t}^{\infty} [h^2(\tau) + h(\tau) n_1(\tau) + h(\tau) n_2(\tau) + n_1(\tau) n_2(\tau)] \, d\tau$$

$$= \int_{t}^{\infty} h^2(\tau) \, d\tau + K(t) \tag{3.3.46}$$

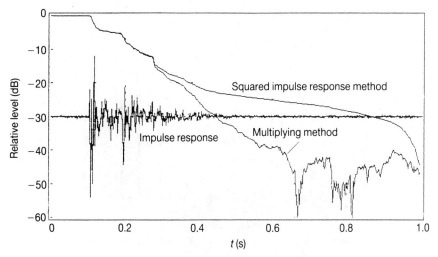

Fig. 3.3.3. Reverberation measurement method by multiplying two impulse response records. Presented by Ono-Sokki.

where $h_1(t) = h(t) + n_1(t)$, $h_2(t) = h(t) + n_2(t)$ and $n_1(t)$ and $n_2(t)$ are the background noise in the measurements. $K(t)$ is either positive or negative and will appear as a fluctuation addition to the first term which gives us the reverberation decay curve. Figure 3.3.3 shows a comparison of the sound decay curves obtained from the multiplied responses and the squared response.

B. *Reverberation Time Calculation*

Examples of the reverberation time in a special room, rather than an ordinary rectangular room, are presented here. This room has a number of scattering obstacles and was designed as a reverberation-variable room at the NTT Human Interface Laboratories in Japan. For some room configurations, the acoustic field can be modelled using the almost-two-dimensional diffuse field theory.

(1) *Reverberation-Variable Room.* Figures 3.3.4a and b show the vertical and horizontal aspects of the room, respectively. The geometric and acoustic parameters are given in Table 3.3.1.

(2) *Reverberation Time in the Room.* Figure 3.3.5 shows measurement results for the reverberation time under three different room conditions. In condition A, all the walls, the floor and the ceiling are reflective. In condition B, all the walls, the floor and the ceiling are absorptive. The reverberation time at 500 Hz is 0.65 s under condition A, and 0.07 s under condition B. The reverberation time under condition C has different frequency characteristics from either A or B. In condition C, all the side walls are reflective as in condition A, while the floor and ceiling are absorptive, as in condition B. The reverberation time becomes longer as the frequency increases, and above 1 kHz the reverberation time for this condition is virtually the same as for condition A.

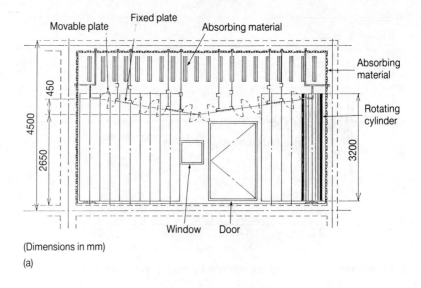

(Dimensions in mm)

(a)

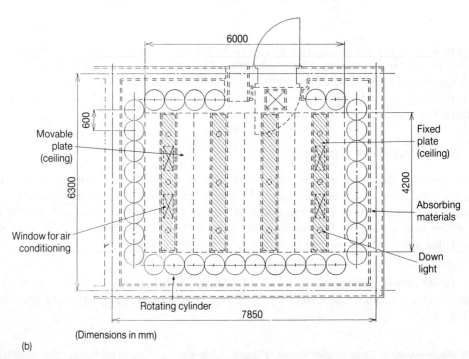

(Dimensions in mm)

(b)

Fig. 3.3.4. Configuration of reverberation-variable room. From Tohyama and Suzuki (1986a).

(3) *Calculation of the Reverberation Time.* Figure 3.3.6 shows the reverberation time as calculated and measured under condition C. The results calculated using the almost-two-dimensional diffuse field theory given by Eq. (3.3.41) are for the most part in good agreement with the measured results. The reverberation time frequency characteristics cannot be predicted using conventional reverberation theories (denoted by ○ and □).

Although the reverberant sound field under condition C was actually produced in the special reverberation-variable room, it could be assumed to be composed mainly of

Table 3.3.1. Geometric and acoustic parameters for the reverberation variable room. From Tohyama and Suzuki (1986a).

Geometric parameters		Average absorption coefficients of reflective side walls, $\tilde{\alpha}_{xy}$	
Section	Figure	Frequency (Hz)	$\tilde{\alpha}_{xy}$
Floor S_{xy}	$25.2\,\text{m}^2\ (=L_xL_y)$	125	0.19
Circumference L_{xy}	$20.4\,\text{m}\ (=2(L_x+L_y))$	250	0.26
Length L_x	$6.0\,\text{m}$	500	0.23
Width L_y	$4.2\,\text{m}$	1000	0.23
Height L_z	$2.875\,\text{m}$	2000	0.20
Volume V	$72.45\,\text{m}^3\ (=L_xL_yL_z)$	4000	0.18
		8000	0.24

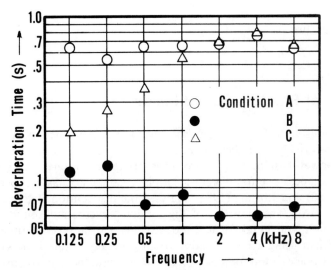

Fig. 3.3.5. Reverberation time in the reverberation-variable room. Condition A: all reflective; condition B: all absorptive; condition C: all the side walls are reflective, the floor and ceiling are absorptive. From Tohyama and Suzuki (1986a).

2-dimensonal waves, parallel to the floor, reflected repeatedly by only the side walls. Other types of 2-dimensional waves do not contribute significantly since the reflective side walls are perpendicular to the floor, and both the floor and ceiling are highly absorptive. Furthermore, 1-dimensional axial waves are not significant, because there are no parallel walls in the nonrectangular room and many scattering cylinders are used in the side walls. The absorption coefficients of the side walls needed for the calculation are given in Table 3.3.1; those of the ceiling and floor, $\tilde{\alpha}_z$, are assumed to be 0.8 on average. In the calculations, account is not taken of the frequency dependency of the absorption coefficients of the ceiling and floor, in order to emphasize the frequency characteristics of the reverberation time derived from the almost-two-dimensional diffuse theory.

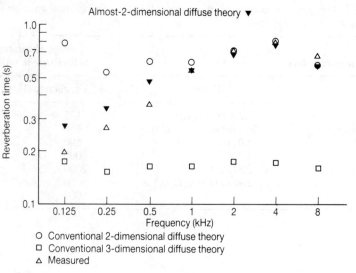

Almost-2-dimensional diffuse theory ▼

O Conventional 2-dimensional diffuse theory
□ Conventional 3-dimensional diffuse theory
△ Measured

Fig. 3.3.6. Calculated and measured reverberation time characteristics at condition C. From Tohyama and Suzuki (1986a).

3.4. Statistics of Random Sound Fields

As described previously, the impulse response and the transfer function from a source to a listener are, in principle, predictable in a deterministic way in an enclosed space. In many cases, however, it may be more appropriate to assume that the transfer function sample variances due to different locations create a stochastic processes.

Waterhouse (1968) and Ebeling (1985) analysed the distribution statistics of sound in a reverberation field. The statistics of the random reverberant sound fields is a particularly important issue in taking acoustic measurements in reverberant enclosures. Acoustic measurement using a reverberation room generally requires the sampling of an acoustic field at one or more points in order to estimate the acoustic properties of the entire field. For example, in measuring the power output of a source in a reverberation room, the space average of the energy density of the sound field is estimated from the values of the mean squared pressure at randomly spaced points. The accuracy of such estimation depends on the statistical properties of the sound field.

3.4.1. Potential Energy Distribution in a Random Sound Field

The addition of sine waves having random amplitudes and/or phases frequently occurs in practice. The sound field in a room is a typical example of such random phenomena. The sound field at any point is created by a large number of plane-wave trains whose phases and incidence directions are randomly distributed. Suppose the mean squared pressure (proportional to the potential energy density) is measured at different points in the room, sufficiently far apart for the phases of the wave trains to be statistically independent. We take an arbitrary point in the space. The total sound pressure at the point is then

$$p(t) = p_0 \sum_{i=1}^{n} \cos(\omega t + \alpha_i)$$

$$= p_0 \left[\sum_{i=1}^{n} \cos \alpha_i (\cos \omega t) - \sum_{i=1}^{n} \sin \alpha_i (\sin \omega t) \right]$$

$$= B \cos(\omega t + \phi) \quad \text{(Pa)} \tag{3.4.1}$$

where B and ϕ are functions of p_0 and α_i. The amplitude and phase are random variables of the frequency and the observation point in the field. At this point, there is a set of n phase angles α_i, distributed from 0 to 2π, and a fixed value of the mean squared pressure,

$$\overline{p_1^2} = \tfrac{1}{2} p_0^2 \left[\left(\sum_{i=1}^{n} \cos \alpha_i \right)^2 + \left(\sum_{i=1}^{n} \sin \alpha_i \right)^2 \right]$$

$$= \tfrac{1}{2} p_0^2 (r_1^2 + s_1^2) \quad \text{(Pa}^2) \tag{3.4.2}$$

At another point, there will be another set of phase angles, and another value $\overline{p_2^2}$. Sampling at many well-spaced points gives the distribution of $\overline{p_k^2}$.

The distribution of α_i is flat, as assumed. When α_i takes any value between 0 and 2π, the values of $\cos \alpha_i$ are distributed between the limits ± 1. The distribution $\sum_{i=1}^{n} \cos \alpha_i = r_k$ is not normal, but does approach normality as $n \to \infty$, by the central limit theorem. Thus, for a large number n, we can treat r_k and s_k as normally distributed variables. In other words, the mean squared pressure is also a random variable given by the squared sum of two normally distributed variables that have zero mean and variance $n/2$ and are mutually independent.

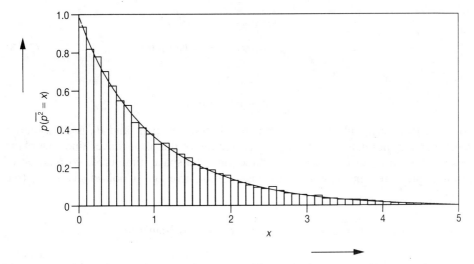

Fig. 3.4.1. Exponential distribution of the potential energy density. The histogram is measured in a reverberation room excited at 5 kHz. From Ebeling (1985).

It is convenient to normalize the expected value of $\overline{p_k^2}$ (ensemble average of $\overline{p_k^2} : \langle \overline{p_k^2} \rangle$) to unity, which corresponds to setting

$$\langle \overline{p_k^2} \rangle = \tfrac{1}{2} n p_0^2 = 1 \quad (\text{Pa}^2) \tag{3.4.3}$$

In this case, the distribution of the sampled value of $\overline{p_i^2} = \overline{p^2}$ has the probability density function

$$p(\overline{p^2} = x) = e^{-x} \tag{3.4.4}$$

and belongs to a family of Gamma distributions. Ebeling (1985) theoretically and experimentally confirmed this distribution as shown in Fig. 3.4.1. The probability density function of $\overline{p^2}$ follows an exponential curve independent of the exciting frequency and the most probable value of $\overline{p^2}$ is zero. In a reverberation room, we can no longer expect a uniform distribution of mean squared sound pressure using the assumption that a sound source radiates a pure sinusoidal wave.

3.4.2. Spatial Correlations for the Sound Pressure Response

We now consider the correlation properties of sound pressure variables between two sampled points in a room. In making reverberation room measurements, we want the measuring points to be far enough apart for the signals to yield statistically independent variables.

The random sound field can be idealized as consisting of a number of plane waves which are uniformly distributed in all directions. The sound pressure at two points r_i and r_j in a local area of the random field can then be written as

$$p_i(\mathbf{r}_i, t_i) = \cos(\omega t_i) \quad (\text{Pa})$$

$$p_j(\mathbf{r}_j, t_j) = \cos(\omega t_j + \phi_j)$$

$$= \cos(\omega t_j - kr \cos \theta_{ij})$$

$$= \cos(\omega(t_i + \tau) - kr \cos \theta_{ij}) \tag{3.4.5}$$

where

$$r = |\mathbf{r}_i - \mathbf{r}_j| \quad (\text{m})$$

$kr \cos \theta_{ij}$ is the phase difference between the observation pair, and θ_{ij} is the random variable corresponding to the propagation direction of the plane wave. Since the plane waves are uniformly distributed in all directions, the space–time cross correlation of the sound pressure can be written as an ensemble average (Cook et al., 1955) (see Section 2.2.3),

$$\langle \overline{p(\mathbf{r}_i, t_i) p(\mathbf{r}_j, t_j)} \rangle = \frac{1}{4} \int_0^\pi \cos(\omega \tau + k\mathbf{r} \cos \theta) \sin \theta \, d\theta$$

$$= \frac{1}{2} \frac{\sin kr}{kr} \cos \omega \tau \tag{3.4.6}$$

Consequently, the normalized space–time cross correlation is

$$R_0(\tau, r) = \frac{\sin kr}{kr} \cos \omega \tau \tag{3.4.7}$$

In making reverberation room measurements, one samples the sound pressure in the field at one or more points and uses the values as a measure of the energy density in the room. To ensure getting independent samples of p^2, it is necessary that the spatial correlation between paired samples is negligibly small. Mean square pressure is defined by

$$\overline{p^2(r_i, t)} = \lim_{T \to \infty} \frac{1}{T} \int_0^T p^2(r_i, t) \, dt \quad (\text{Pa}^2). \tag{3.4.8}$$

Similarly to the approach above, we take the ensemble average of the paired mean square pressure data as

$$\langle \overline{p^2(r_i, t_i)} \, \overline{p^2(r_j, t_j)} \rangle = \lim_{T \to \infty} \frac{1}{T^2} \int_0^T \int_0^T \langle p^2(r_i, t_i)p^2(r_j, t_j) \rangle \, dt_i \, dt_j$$

where

$$\langle p^2(r_i, t_i)p^2(r_j, t_j) \rangle = \langle p^2(r_i, t_i) \rangle \langle p^2(r_j, t_j) \rangle + 2\langle p(r_i, t_i)p(r_j, t_j) \rangle^2$$

and

$$\langle \overline{p^2(r_i, t_1)} \rangle = \langle \overline{p^2(r_j t_2)} \rangle = \overline{p^2} \tag{3.4.9}$$

We have the cross correlation of the mean squared pressure data (Chu, 1982; Lubman, 1969)

$$R(r) = 2 \lim_{T \to \infty} \frac{1}{T^2} \int_0^T \int_0^T \langle p(r_i, t_i)p(r_j, t_j) \rangle^2 \, dt_i \, dt_j$$

$$= \left(\frac{1}{2} \frac{\sin kr}{kr} \right)^2 \tag{3.4.10}$$

The normalized cross correlation for the mean squared pressure data sampled at two locations with separation r is therefore

$$R_0(r) = \left(\frac{\sin kr}{kr} \right)^2 \tag{3.4.11}$$

3.4.3. Distribution of Squared Pressure Averaged at N Points

Now let us consider the distribution of the mean squared pressure as averaged at N well-separated points. That is, we select points with random separations, greater than $\lambda/2$ where λ is the wavelength of the sound. We can expect that this averaging will decrease the variance.

Suppose that the value of the mean squared pressure measured at one point is $\overline{p_k^2}$. The sum of the pressures measured at different points is then (see Eq. 3.4.2)

$$\overline{p^2} = \sum_{k=1}^{N} \overline{p_k^2} = \tfrac{1}{2} p_0^2 \sum_{k=1}^{N} (r_k^2 + s_k^2) \quad (\text{Pa}^2) \tag{3.4.12}$$

If we can assume that r_k and s_k are statistically independent, both $\overline{p_k^2}$ and $\overline{p^2}$ are distributed following Gamma distributions. This follows because the sum of Eq. (3.4.12) contains N terms like Eq. (3.4.2) and the Gamma distribution reproduces itself by composition. The Gamma distribution function has the probability density function (Waterhouse, 1968), as shown in Eq. (3.3.18),

$$\gamma(x) = [(N-1)!]^{-1} N^N e^{-Nx} x^{N-1} \tag{3.4.13}$$

where the mean value is unity and the variance is $1/N$. The variance of the mean squared pressure data therefore decreases inversely as the number of sampling points increases.

3.4.4. Spatial Coherence for the Sound Pressure Response

Another important application of response statistics is determining the distance of coherence. The coherence function between two signals $x(t)$ and $y(t)$ is defined as (Bendat and Piersol, 1971) (see Section 2.2.5)

$$\gamma_{xy}(\omega)^2 = \frac{|G_{xy}(\omega)|^2}{G_{xx}(\omega)G_{yy}(\omega)} \tag{3.4.14}$$

where $G_{xx}(\omega)$ and $G_{yy}(\omega)$ denote the power spectra of $x(t)$ and $y(t)$, respectively, and $G_{xy}(\omega)$ is the cross spectrum between them. If we assume that the response forms a diffuse wave field, it is possible to compute the coherence as a function of both range and frequency.

Suppose that r is the separation between the observation points for two signals and

$$x(t) = A e^{-jkx_1 + j\omega t} \qquad y(t) = A e^{-jkx_2 + j\omega t} \tag{3.4.15}$$

The coherence function becomes

$$\gamma_{xy}^2 = J_0^2(kr) \tag{3.4.16a}$$

in a 2-dimensional diffuse field, and

$$\gamma_{xy}^2 = \left(\frac{\sin kr}{kr}\right)^2 \tag{3.4.16b}$$

in a 3-dimensional diffuse field. The coherence given by these equations depends on the frequency, and gives the frequency dependence for a fixed separation in a diffuse field.

Figure 3.4.2a shows the measured results of the coherence for two dimensions (an elastic plate) and three dimensions (a small box) by Liu and Lyon (1991) and comparisons with the calculated values from Eq. (3.4.16). Sketches of these experimental systems are shown in Fig. 3.4.2b. Each system is excited by a band noise, and the coherence is averaged over a group of paired measurement locations. Each pair of locations is at the same separation distance, but the positions and orientations of the

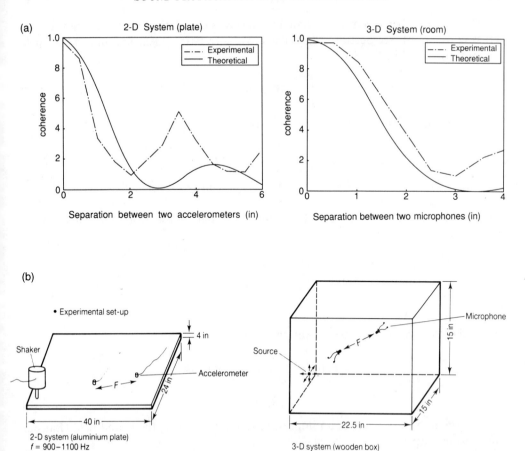

Fig. 3.4.2. Coherence function in a diffuse field. (a) A comparison of the theoretical diffuse field coherence and the experimental results for the finite systems shows a tendency for partial coherence to be enhanced in the finite (real) system. (b) The 2-dimensional system is an aluminium plate and the 3-dimensional system is a wooden box cavity. The frequency bands are chosen so that the wavelength in these systems is about the same. The noise bandwidth is 200 Hz in both cases. Presented by Liu and Lyon (1991).

paired locations are randomly selected. We see in Fig. 3.4.2 that the general form of the spatial coherence follows the theoretical behaviour, but there are significant differences in detail.

3.4.5. Wavenumber Statistics for Random Sound Fields

A random sound field may be described using the random-phase plane waves. We assume these plane waves are uniformly distributed in all directions in the diffuse field. This distribution function for the directions of the approaching plane wave can be formulated by using the wavenumber spectrum (Lubman, 1974; Tohyama et al, 1977; Ebeling, 1985).

The wavenumber spectrum is related to the cross correlation coefficient (normalized cross correlation) (see Section 2.2.3). The cross correlation coefficient in a diffuse field is given by ($\tau = 0$ in Eq. 3.4.7)

$$R_0(r) = \int_0^{+\infty} W(k) \cos kr \, dk = \frac{\sin k_0 r}{k_0 r} \qquad (3.4.17)$$

where $W(k)$ denotes the wavenumber spectrum,

$$W(k) = \begin{cases} \dfrac{1}{k_0}, & 0 < k < k_0 \\ 0 & \text{elsewhere} \end{cases} \qquad (3.4.18)$$

in a 3-dimensional diffuse field. In a 2-dimensional diffuse field, the wavenumber spectrum is

$$W(k) = \begin{cases} \dfrac{2}{\pi} \dfrac{1}{\sqrt{k_0^2 - k^2}} & 0 < k < k_0 \\ 0 & \text{elsewhere} \end{cases} \qquad (3.4.19)$$

Thus, the cross correlation coefficient becomes

$$R_0(r) = \int_0^{+\infty} W(k) \cos kr \, dk$$

$$= J_0(k_0 r) \qquad (3.4.20)$$

The wavenumber spectrum is an important tool for assessing the sound field (Bakster and Morfey, 1986).

3.5. Transfer Function Statistics

The transfer function (TF) between a pair of source and listening positions is defined as a Laplace transform of the impulse response (see Chapter 2). While a TF is determined by a wave equation in the sound space, it generally has numerous poles and zeros. As we described in the previous section, it may be more appropriate to assume that the TF samples due to different locations make an ensemble of stochastic processes.

Perhaps Wente (1935) made the earliest scientific report in which he considered the transmission irregularity of the TFs in a room. Schroeder (1954a, b) formulated the TF as a stochastic process and analysed its stochastic properties under the high modal overlap condition. Lyon (1969) explicitly introduced the modal overlap concept into TF analysis and explored statistical TF analysis under the low modal overlap condition based on Poisson statistics with random occurrences of poles.

3.5.1. Transfer Function in a Room

The TF in an enclosed space with a simple harmonic source (assuming time dependency of $\exp(j\omega t)$) at X_s and an observer at X_0 is given by a modal expansion form (Morse and Bolt, 1944; Kuttruff, 1990),

$$H(\omega) = \text{const} \sum_M \frac{\Psi_M(x_s)\Psi_M(x_0)}{\omega^2 - \Omega_M^2} \quad (\text{Pa s/m}^3), \tag{3.5.1}$$

where Ψ_M are eigenfunctions (normal mode or mode shapes), $\omega = 2\pi f$, f is the frequency in Hertz, and Ω_M are the eigenvalues of the space under the boundary condition. Generally, both Ψ_M and Ω_M are complex functions. If the walls are hard and their impedances are frequency independent, the eigenvalues and functions are real and independent of the frequency.

An eigenfunction has both loops and nodes. In a 1-dimensional system, the nodes are points; in a 2- or 3-dimensional system, however, they are lines or surfaces that divide the enclosed space. If we imagine a nondissipative system, the zeros of the TF create the nodal lines (or surfaces). These nodal lines are not observed in an actual vibrating system. If there are such continuous nodal lines in the space, any source position is surrounded by the closed nodal line, that is, no energy is distributed into the space. The zeros of the TF must therefore be discretely distributed, like a cloud in a vibrating system space, where some modal overlap is expected due to modest damping (Lyon, 1992). Lyon (1991) called the distribution of zeros of TFs a cloud of zeros.

3.5.2. Poles and Zeros of Transfer Functions

The TF can be rewritten as a ratio of polynomials, which are factored according to their roots (Lyon 1983),

$$H(\omega) = K \frac{(\omega - \omega_a)(\omega - \omega_b) \cdots}{(\omega - \omega_1)(\omega - \omega_2) \cdots} \quad (\text{Pa s/m}^3) \tag{3.5.2}$$

where ω_1 and ω_2 are the poles (the singularities of the TF), ω_a and ω_b are the zeros of the function, and K is a constant. The poles, which are determined by the eigenvalues, are independent of the source and observation point locations. The zeros, however, are dependent on their locations.

Since $\exp(j\omega t)$ time dependency is assumed, all the poles must be in the upper half-plane in the complex frequency plane because of causality in the system. The nth pole, which gives the nth frequency of the free vibration (nth natural frequency), is at $\omega = \omega_n + j\delta_n$. This complex frequency plane corresponds to the jp^*-plane which was introduced in Section 2.3. Here * denotes taking the conjugate value. That is, the left-half plane in the p-plane corresponds to the upper-half plane of this complex frequency plane.

For simplicity we assume that the reverberation time $T_R = 6.9/\delta_n$ is the same for all modes ($\delta_n = \delta_0$). The poles lie along a straight line (pole-line) at a distance $\delta_0 = 6.9/T_R$ above the real frequency line (Fig. 3.5.1). The TF is symmetric in the complex frequency plane with respect to the pole-line when the eigenfunctions are real (Tohyama and Lyon, 1989a).

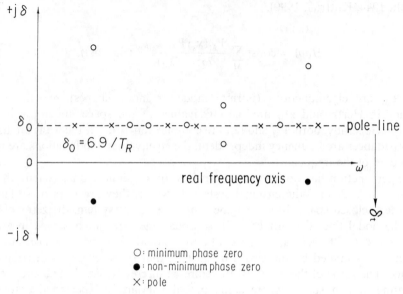

O: minimum phase zero
● : non-minimum phase zero
X: pole

Fig. 3.5.1. Poles and zeros in the complex frequency plane. From Tohyama et al. (1991).

3.5.3. Modal Density and Modal Overlap

The number of eigenvalues in a unit frequency interval is called the modal density (Morse and Bolt, 1944). The room eigenfunctions (modes) are classified into three groups, which are composed of oblique, tangential, and axial waves. The modal density for each is

$$n_{pob} \approx \frac{V\omega^2}{2\pi^2 c^3} \quad \text{(oblique waves)} \tag{3.5.3a}$$

$$n_{ptan} \approx \frac{S\omega}{2\pi c^2} \quad \text{(tangential waves), and} \tag{3.5.3b}$$

$$n_{pax} \approx \frac{L}{\pi c} \quad \text{(axial waves),} \tag{3.5.3c}$$

where V denotes the volume (m^3), S denotes the area (m^2) of the field in which the tangential modes of interest are excited, and L is the length (m) of the axial wave field.

The modal density of the room rapidly increases as the frequency increases. If we assume modest damping of the sound field, then many modes are excited simultaneously by a single frequency. The number of such simultaneously excited modes is called the modal overlap. The modal overlap is defined as the product of the modal bandwidth and the modal density (Lyon, 1969),

$$M = Bn_p(\omega) = \pi\delta_0 n_p(\omega) \simeq \pi\delta_0 n_{pob}(\omega) \tag{3.5.4a}$$

where B is the modal bandwidth,

$$B = \frac{1}{H_{0max}^2} \int_0^\infty |H_0(\omega)|^2 \, d\omega = \pi\delta_0 \quad \text{(rad/s)} \tag{3.5.4b}$$

H_0 is the frequency response of the single modal function, and H_{0max} is the maximum of the absolute value of the response.

3.5.4. Distribution of Eigenvalues (Eigenfrequencies)

The poles lie along the pole-line. The eigenvalues of a rectangular room (L_x, L_y, L_z) with rigid boundaries are given by

$$\omega_{lmn} = (c\pi/L)\sqrt{(al)^2 + (bm)^2 + (cn)^2} \quad (rad/s) \qquad (3.5.5)$$

where l, m, and n are integer numbers, the length of the sides (m) are $L_x = xL$, $L_y = yL$ and $L_x = zL$, and x, y and z are the ratios of the lengths. Bolt (1947) introduced a Poisson model that produces an exponential distribution of the eigenvalue spacings. Figure 3.5.2 shows the eigenvalue spacing histogram of the oblique waves in a rectangular room.

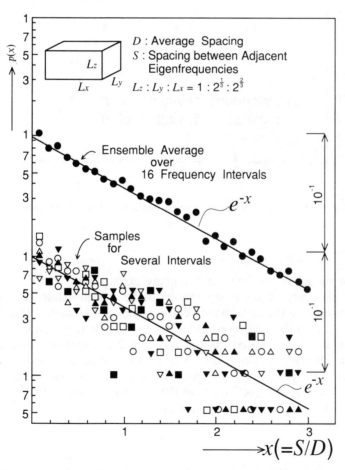

Fig. 3.5.2. Eigenvalue spacing histogram of a rectangular room. From Tohyama et al. (1992c).

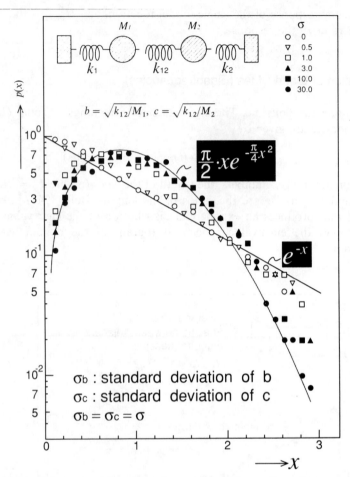

Fig. 3.5.3. Eigenvalue spacing histogram of coupled oscillators. From Tohyama et al. (1992c).

The Poisson model is not suitable for many actual cases, as Schroeder (1989) suggested. The tendency of small separations to be less probable can perhaps be explained by considering the perturbation effects (Lyon, 1969; Davy, 1990). The effects of the perturbations on the spacing statistics can be illustrated by coupled mechanical oscillators. Wigner (1965) showed theoretically that 2×2 matrices have eigenvalue separation that follows a Rayleigh distribution rather than a Poisson distribution. Referring to Fig. 3.5.3, the masses and spring constants of the two oscillators are M_1 and M_2, and K_1 and K_2, respectively, and the coupling spring constant is K_{12}. The 'eigenmatrix E' for the eigenvalues is written as

$$(E) = (E_0) + (E_p) = \begin{pmatrix} \omega_1^2 & 0 \\ 0 & \omega_2^2 \end{pmatrix} + \begin{pmatrix} b^2 & -b^2 \\ -c^2 & c^2 \end{pmatrix} \qquad (3.5.6)$$

where $\omega_1 = \sqrt{K_1/M_1}$, $\omega_2 = \sqrt{K_2/M_2}$, $b = \sqrt{K_{12}/M_1}$ and $c = \sqrt{K_{12}/M_2}$. In Eq. (3.5.6), matrix E_0 has the eigenvalues of two independent oscillators and matrix E_p represents the coupling effects.

The eigenvalues of the coupled oscillators are given by the positive square roots of the eigenvalues λ_1 and λ_2 of the matrix E. The perturbation caused in an actual acoustic space like an irregularly shaped room results in random coupling between adjacent modes (Morse and Bolt, 1944; Koyasu and Sato, 1957). We can assume that this modal coupling varies randomly as the frequency changes. Following Wigner's (1965) random matrix theory (Mehta, 1991), suppose that the coupling parameters b and c of matrix E_p are mutually independent Gaussian variables with zero mean value and a standard deviation of σ. The spacing of the two eigenvalues, $S = \Delta\omega = \sqrt{\lambda_1} - \sqrt{\lambda_2}$, can be expressed by

$$S^2 = (\Delta\omega)^2$$
$$= \omega_1^2 + \omega_2^2 + b^2 + c^2 - 2\sqrt{(\omega_1^2 + b^2)(\omega_2^2 + c^2) - b^2 c^2} \approx b^2 + c^2 \qquad (3.5.7)$$

as σ becomes much larger than ω_1 and ω_2. The random variable $\Delta\omega$ is the positive square root of the squared sum of the two independent Gaussian variables. Consequently, the spacing between two eigenvalues follows a Rayleigh distribution as the perturbation becomes large. Figure 3.5.3 illustrates the transition of the spacing histograms from a Poisson distribution to a Rayleigh distribution. We randomly produced 5000 pairs of eigenfrequencies with spacings that follow a Poisson distribution with a zero coupling effect. From the illustrated results, we can surmise that the Rayleigh distribution of the eigenfrequency spacing is produced by the randomly coupled wave modes (Lyon, 1969; Wigner 1965; Davy, 1990).

3.5.5. Distribution of Zeros

While the poles are distributed along the pole-line following a Rayleigh distribution, the zeros are distributed 2-dimensionally throughout the complex frequency plane. Zeros in the TF of a slightly damped system are usually located at the points where the contributions from the two adjacent poles cancel each other. The distribution of zeros was explored by Lyon (1983, 1984, 1987) who emphasized that the phase property of the TF is quite important for sound field and vibration control. The phase can be estimated from both the poles and zeros in the complex plane.

A. Low Modal Overlap Case

We introduce an approximate formula by considering a test frequency between two adjacent poles. The primary part of the formula dominates the TF in the modal expansion, while the remainder produce a fairly slow remainder function R. This remainder function corresponds to the sum of the contributions from poles excluding the two adjacent poles (Tohyama and Lyon, 1989a).

If the system damping is very small, the TF can be approximately written as

$$H(\omega) \approx \frac{A}{\omega - \omega_1} + \frac{B}{\omega - \omega_2} + R \qquad (3.5.8)$$

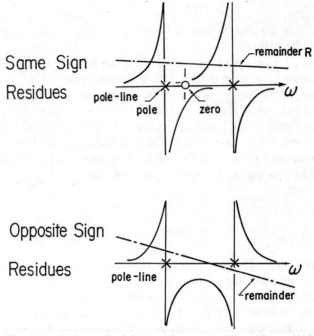

Fig. 3.5.4. Poles and minimum phase zeros. From Lyon (1987).

A minimum phase zero (see Section 2.2) is produced between the two adjacent poles when they have residues of the same sign. Such a zero is located on the line connecting the poles (the pole-line), as shown in Fig. 3.5.4. If the TF has opposite sign residues, the occurrence of zeros will depend on the remainder function R. There are two cases of zeros that are produced from opposite sign residues and the remainder function as shown in Fig. 3.5.5. One has double zeros between two adjacent poles, both of which are minimum phase because they are located on the pole-line. The other has a conjugate pair of zeros in the complex frequency plane. Those zeros are located symmetrically with respect to the pole-line. One of this pair of zeros can be of nonminimum phase.

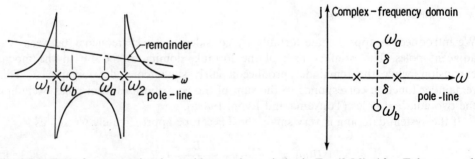

Fig. 3.5.5. Zeros from opposite sign residues and remainder in Eq. (3.5.8). After Tohyama and Lyon (1989a).

B. *High Modal Overlap Case*

We again take a TF with a modal expansion form of Eq. (3.5.1). When the source and observer are at the same location, all the residues are positive. The poles therefore interlace with zeros and the numbers of poles and zeros are equal. As the source and receiver move apart, the zeros migrate (poles do not move). Some move above the line and an equal number of zeros move symmetrically below the line. Other zeros remain on the line (Lyon, 1983, 1984, 1987).

Consider an expansion form in a 2-dimensional rectangular system as an example,

$$H(\omega) = \text{const} \sum_M \frac{(-1)^{l+m}}{\omega^2 - \omega_M^2} \quad (\text{Pa s/m}^3) \tag{3.5.9}$$

where the source is located at $(0, 0)$ and the receiving point at (L_x, L_y). If we rank the terms in the sum of Eq. (3.5.9) according to increasing pole values, we cannot tell analytically whether the sum $l + m$ will change its parity and, therefore, change the residue sign. Since the eigenvalues are determined from the modal lattice shown in Fig. 3.5.6, the sum $l + m$ will as likely change parity as not between successive poles. The number of zeros on the pole-line is about half the number of poles, since we can assume equal probability for the same or opposite sign for adjacent pole residues.

If the eigenfunctions are real, and the pole line is as shown in Fig. 3.5.1, then by symmetry the remaining zeros are evenly distributed above and below the pole-line, which means that 1/4 of them are below the pole-line. The maximum number of zeros below the real frequency axis is therefore $N_p/4$, where N_p denotes the total number of poles, $N_p = N_z^+ + N_z^-$, and N_z^+ and N_z^- denote the number of nonminimum and minimum phase zeros, respectively. The number of nonminimum phase zeros is reduced, as $\delta_0 = 6.9/T_R$ increases.

We will consider the TF for a reverberant field to be a random process in the frequency domain (Lyon, 1969; Schroeder 1954a, b, 1962). The TF $H(f)$ can be

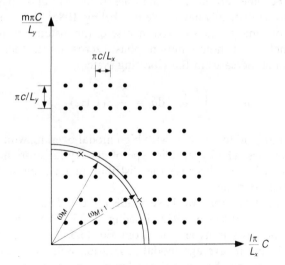

Fig. 3.5.6. Wavenumber lattice for a 2-dimensional room showing how the sequence of eigenvalues is generated and how the modal index sum $l + m$ tends to have a random parity. After Lyon (1987).

written as

$$H(f) = H_r(f) + jH_i(f) \tag{3.5.10}$$

where f is the frequency in Hertz. The inverse Fourier transform of $H(f)$ is the impulse response $h(t)$ between the two points. The inverse Fourier transform of the real part $H_r(f)$ is $h(t)/2$, and its power, $w(t)$, is $h^2(t)/4$. The ensemble average of $h^2(t)$ is

$$4W(t) = 4\langle w(t)\rangle = \langle h^2(t)\rangle = E_0 e^{-t/\tau} \quad (t > 0), \tag{3.5.11}$$

where $\tau = T_R/13.8$ and T_R is the reverberation time of the field (Schroeder, 1962).

The expected number of zero crossings per unit increase in f is given by (Rice, 1945; Soong, 1973)

$$n_{rz} = \left[\frac{\int_0^\infty t^2 W(t)\, dt}{\int_0^\infty W(t)\, dt} \right]^{1/2} \quad (1/\text{Hz}) \tag{3.5.12}$$

Introducing Eq. (3.5.11) into Eq. (3.5.12), the expected number of zeros per Hertz in the real part of the $H(f)$ is

$$n_{rz} = \frac{\sqrt{2}}{\delta_t} \quad (1/\text{Hz}) \tag{3.5.13}$$

where δ_t is the distance from the pole-line in the complex (angular) frequency domain as shown in Fig. 3.5.1. The function $H(f)$ has zeros at frequencies where both the real and imaginary parts become zero at the same time. If the real and imaginary parts of $H(f)$ can be considered statistically independent (Ebeling, 1985), then the density of TF zeros should be proportional to the squared inverse of the distance from the pole-line. The distribution function of nonminimum phase zeros must, therefore, decrease inversely as the damping increases in the vibrating system,

$$n_z \sim 2 \int_{\delta 0}^\infty \delta_t^{-2}\, d\delta_t = \frac{2}{\delta_0} \quad (1/\text{Hz}) \tag{3.5.14}$$

This result is expected to hold for systems with high modal overlap, which is the normal situation in room acoustics. The hyperbolic type of density function, however, cannot give the total number of zeros, because the distribution function will diverge as it approaches close to the pole-line.

The statistical assumptions that we have made for the TFs hold well with large modal overlap (Schroeder, 1954a, b, 1962; Ebeling, 1985). We can expect that the distribution function of zeros is hyperbolic from Eq. (3.5.14) when the ratio of the distance from the pole-line to average modal separation is large, because the modal overlap is large in that region. The modal overlap decreases as the distance between the real frequency axis and the pole-line decreases. The density function of zeros must depend on the ratio $\beta = n_p\delta_t = \delta_t/D_p$. Following Eq. (3.5.4a), the modal overlap at $\delta_0 = \delta_t$ is

$$M = \pi n_p(\omega)\delta_t = \pi\beta \tag{3.5.15}$$

The density of zeros off the pole-line must therefore collapse into the pole-line as the modal density increases (Lyon, 1983, 1984). A probability density function of zeros that has the required form for large β is integrable,

$$p(\beta) = \frac{2\sqrt{\gamma/\pi}}{1 + \gamma\beta^2} \tag{3.5.16}$$

where γ is a shape factor. The total number of zeros up to angular frequency ω is

$$N_z^+(\delta_0, \omega) = \int_0^\omega \tfrac{1}{4} n_p(\omega)\,d\omega \int_{\beta_0}^\infty p(\beta)\,d\beta$$

$$= N_z^+(0, \omega)[1 - \Gamma(\zeta_l)] \tag{3.5.17}$$

where $n_p(\omega) = D\omega^2$, $D = V/2\pi^2 c^3$, $\zeta_l = [n_p(\omega)\delta_0\gamma^{1/2}]^{1/2}$, $\beta_0 = \delta_0/D_p = n_p\delta_0$, and

$$\Gamma(\zeta_l) = \frac{6}{\pi\zeta_l^3} \int_0^{\zeta_l} \zeta^2 \tan^{-1}(\zeta^2)\,d\zeta \tag{3.5.18}$$

Equation (3.5.18) is graphed in Fig. 3.5.7.

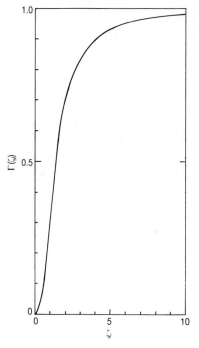

Fig. 3.5.7. Distribution variable for zeros. From Tohyama et al. (1991).

3.6. Statistical Phase Analysis (SPA)

The control of acoustical systems or machinery noise requires signal processing methods that take into account the complexity and variability of their TFs. A certain stable characteristic in the statistical properties of these TFs, which may be represented by thousands of poles and zeros, are desirable in order to reduce the number of parameters needed to describe them. This reduction of parameters is very helpful for controlling machinery noise and sound fields. Lyon has been exploring statistical phase analysis (SPA) from this point of view.

3.6.1. Propagation and Reverberation Phases

Lyon (1983, 1984) classified the phase in a sound space into propagation and reverberation phases.

A. Propagation Phase

The growth of phase between a source and the observation locations in a 1-dimensional system can be visualized. As shown in Fig. 3.6.1, suppose x_0 and x_s are separated by m nodes so that

$$\text{int}[(k/\pi)(x_0 - x_s)] = m \tag{3.6.1}$$

where int[ ] means taking an integer derived by truncation and k denotes the wavenumber (1/m). Even the mode number increases, these locations stay in phase until m increases by unity. If m increases unity, the phase between observation and source locations increases by $-\pi$. The phase trend in this case, where the number of nodes increases in an orderly fashion between source and receiver, is

$$\Phi = -k(x_0 - x_s) \quad \text{(rad)} \tag{3.6.2}$$

Thus, the accumulated phase of a one-dimensional system is just the same as the phase delay due to the direct wave propagation from source to observer. We call this the propagation phase.

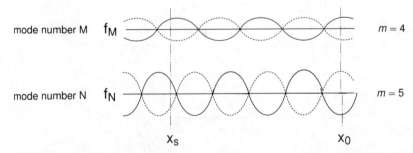

Fig. 3.6.1. Standing wave pattern in a 1-dimensional system showing phase advance as nodal pattern changes. From Lyon (1984).

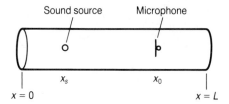

Fig. 3.6.2. Acoustic pipe. Redrawn from Lyon (1987).

B. *Reverberation Phase*

Consider the acoustical pipe in Fig. 3.6.2, which has a volume velocity source at location x_s and a microphone at x_0. The TF given by Eq. (3.5.1) is written in a closed form for the 1-dimensional case as (Lyon, 1983, 1984, 1987)

$$H(\omega) = \text{const } \frac{\cos kx_s \cos k(L - x)}{\sin kL} \qquad (3.6.3)$$

where L denotes the length of the 1-dimensional system. System resonances occur when $k = m\pi/L$, and zeros occur when the distance between the source or the observation points and the ends of the pipe are odd multiples of a quarter-wavelength of sound. That is,

$$k = \frac{m\pi}{L} \qquad \text{for poles}$$

$$k = \frac{(n + \frac{1}{2})\pi}{x_s} \qquad \text{for zeros}$$

$$k = \frac{(r + \frac{1}{2})\pi}{L - x_0} \qquad \text{for zeros} \qquad (3.6.4)$$

where m, n and r are integers. All the poles and zeros are located in the upper half of the complex frequency plane, assuming modest damping.

The 2- or 3-dimensional cases are not so orderly, however, as the 1-dimensional case. In these cases, the accumulated phase can be determined by both the poles and zeros (Fig. 3.6.3). Minimum phase zeros cancel the effects of the poles on the TF phase, while each nonminimum phase zero and each pole increase the phase lag by π. If there are N_p poles, and if there are N_z^- and N_z^+ zeros in the upper and lower half-planes respectively, the accumulated phase is

$$\Phi = -\pi(N_p + N_z^+ - N_z^-) = -2\pi N_z^+ \quad \text{(rad)}, \qquad (3.6.5)$$

where $N_p = N_z^+ + N_z^-$. In these fields, since there is a large number of zeros in the lower half-plane, the accumulated phase estimated by Eq. (3.6.5), which we call the reverberation phase, is much greater than the propagation phase in a 1-dimensional system

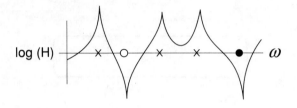

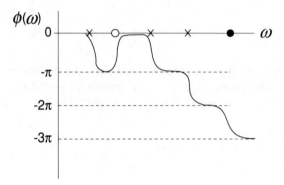

○: minimum phase zero
●: non-minimum phase zero
×: pole

Fig. 3.6.3. Poles, zeros, and accumulated phase. Redrawn from Lyon (1987).

which is expressed by

$$\Phi = -\pi(N_p - N_z^-) = -\pi\left\{\frac{k}{\pi/L} - \frac{k}{\pi/(L - x_o)} - \frac{k}{\pi/x_s}\right\}$$

$$= -k(x_o - x_s) \tag{3.6.6}$$

The TF given by Eq. (3.5.1) includes both the propagation and reverberation phase. When the distance between the source and observation points is zero, the poles interlace with the zeros. As source and receiver move apart, the zeros migrate. If we place a 'barrier' on the pole-line at some frequency, we can count the zeros that migrate across this barrier. The number of such zeros that cross the barrier leaves some 'uncancelled poles' owing to the imbalance in the number of poles and zeros. Consequently, an accumulated phase is produced as the observation point moves away from the source. If we set the barrier at a sufficiently high frequency, the accumulated phase due to the zero migration becomes the propagation phase (Lyon, 1992).

While the zero migration of the minimum phase zeros produces the propagation phase, the nonminimum phase zeros generate the reverberation phase delay between

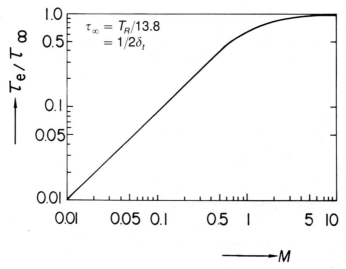

Fig. 3.6.4. Group delay and modal overlap M. From Tohyama et al. (1991).

source and observer. Therefore, following Eqs. (3.5.16) and (3.5.17), the reverberation phase is $\phi_{rev} = -2\pi N_z^+$ and the contribution to ϕ_{rev} in $d\omega$ is

$$d\phi_{rev} = -n_p(\omega)[(\pi/2) - \tan^{-1}(\beta\sqrt{\gamma})]\,d\omega \quad (\text{rad}) \qquad (3.6.7)$$

where $\beta = n_p\delta_t$ and the group delay is (β or M large)

$$\tau_e = -\frac{d\phi_{rev}}{d\omega}$$

$$= \tau_\infty \frac{2}{\sqrt{\gamma}}\left[1 - \frac{1}{3}\left(\frac{\sqrt{\gamma M}}{\pi}\right)^{-2} + \frac{1}{5}\left(\frac{\sqrt{\gamma M}}{\pi}\right)^{-4} - \frac{1}{7}\left(\frac{\sqrt{\gamma M}}{\pi}\right)^{-6} + \cdots\right] \quad (\text{s}) \qquad (3.6.8)$$

and $\tau_\infty = 1/(2\delta_t) = T_R/13.8$ (s). The ratio τ_e/τ_∞ is graphed in Fig. 3.6.4 as a function of the modal overlap M. Since this should be equal to unity for high modal overlap (Schroeder and Kuttruff, 1991), we conclude that $\gamma = 4$ (Cauchy distribution).

3.6.2. Group Delay Function for a TF

The poles have a Poisson-like distribution on the pole-line; the zeros, however, lie 2-dimensionally throughout the complex frequency plane. The phase behaviour can be thought of as a stochastic process due to random occurrences of the poles and zeros. The accumulated phase has large variability from the theoretical trend of the reverberation phase by Eq. (3.6.5) (Lyon, 1987). We will now describe the fluctuations of the TF phase using group delay.

A. Group Delay Random Process

We start by taking the general form of the TF as written in Eq. (3.5.2). The poles and zeros form a pattern in the complex frequency plane as shown by Fig. 3.6.5a. Taking the natural log,

$$\ln |H(\omega)| = \ln |K| + \ln|\omega - \omega_a| + \cdots - \ln|\omega - \omega_1| - \cdots \tag{3.6.9}$$

$$\phi = \arg(\omega - \omega_a) + \arg(\omega - \omega_b) + \cdots - \arg(\omega - \omega_1) - \arg(\omega - \omega_2) - \cdots$$

$$= \sum_m a_m \phi_m \tag{3.6.10}$$

where $a_m = \pm 1$ and $\omega_m = \omega_m^r + j\delta_m$ (poles and zeros of H) and

$$\arg(\omega - \omega_m) = \phi_m = -\tan^{-1} \frac{\delta_m}{\omega - \omega_m^r} \tag{3.6.11}$$

the group delay is

$$\tau(\omega) = -\frac{d\phi}{d\omega} = -\sum_m a_m \frac{d}{d\omega}\left(-\tan^{-1}\frac{\delta_m}{\omega - \omega_m^r}\right)$$

$$= -\sum_m \frac{a_m \delta_m}{(\omega - \omega_m^r)^2 + \delta_m^2} = -\sum_m \left(\frac{a_m}{\delta_m}\frac{1}{\xi_m^2 + 1}\right) = -\sum_m \tau_m \tag{3.6.12}$$

where $\xi_m = (\omega - \omega_m^r)/\delta_m$. This equation represents a summation of 'pulse trains' on a frequency scale due to all the singularities (Lyon, 1969). It can also be rewritten as an integration of group delay pulse-trains based on the density of the singularities of the δ-strip (hatched area in Fig. 3.6.5b), where $\delta_m = \delta$. Since the pulse shape depends only on δ, all the pulses of a pulse train due to the singularities in a δ-strip have the same shape.

The strip off the pole-line has a number of singularities per unit frequency interval of

$$dn_{\text{zero-cloud}}(\omega) = \frac{n_p(\omega)}{4} p(\beta)\, d\beta \tag{3.6.13}$$

where $p(\beta) = (4/\pi)/(1 + 4\beta^2)$ ($\gamma = 4$ in Eq. (3.5.16): Cauchy distribution) is the density of the cloud of zeros for $0 < \beta < \infty$. On the pole-line itself,

$$dn_{\text{pole-line}}(\omega) = \tfrac{3}{2}\, n_p(\omega)\delta(\beta - \beta_0)\, d\beta \tag{3.6.14}$$

since there are n_p singularities due to poles ($a_m = -1$) and $(1/2)n_p$ singularities ($a_m = +1$) due to the zeros on the pole-line, and $\delta(\beta)$ is delta-function of β. They are shown in Fig. 3.6.5a. The values of a_m are determined from the steps in phase for the original TF as modeled in Fig. 3.6.3.

B. Mean Value of Group Delay

If we assume that the frequency separation of the singularities in a δ-strip has a Poisson distribution (Lyon, 1969), the mean value of τ contributed by any strip of singularities

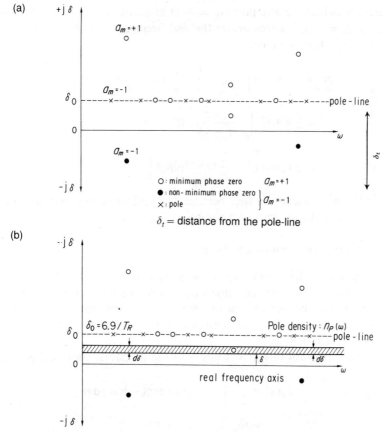

Fig. 3.6.5. Pole/zero locations and a δ strip for the group delay singularities. (a) Signs for the singularities in the group delay. (b) Poles, zeros and δ strip. From Tohyama et al. (1994).

in $d\delta$ (or $d\beta = n_p d\delta = d\delta/D_p$, $\beta = n_p \delta$) is

$$dm_\tau = -dn(\omega) \left\langle \int_{-\infty}^{+\infty} \tau_m(\omega)\, d\omega \right\rangle_m$$

$$\left\langle \int_{-\infty}^{+\infty} \tau_m(\omega)\, d\omega \right\rangle_m = \langle a_m \rangle_m \frac{1}{\delta} \int_{-\infty}^{+\infty} \frac{1}{\xi^2 + 1}\, d\omega$$

$$= \langle a_m \rangle_m \pi$$

$$dm_\tau = -\pi \langle a_m \rangle_m dn(\omega) \tag{3.6.15}$$

where $\langle * \rangle_m$ denotes the average in the frequency interval of interest. Combining pole-line and zero-cloud distribution singularities, we get

$$m_\tau = -\pi n_p(\omega) \int_{-\infty}^{+\infty} \frac{3}{2} \langle a_{m,\text{pole-line}} \rangle_m \delta(\beta - \beta_0)\, d\beta - \pi n_p(\omega) \int_{-\infty}^{+\infty} \frac{1}{4} \langle a_{m,\text{zero cloud}} \rangle_m P(|\beta|)\, d\beta \tag{3.6.16}$$

Since there are twice as many poles ($a = -1$) as zeros ($a = +1$) on the pole line, we get $\langle a_{m,\,\text{pole-line}} \rangle_m = -1/3$. Zeros above the real frequency axis have $a = +1$ and those below have $a = -1$, so we have

$$m_\tau = -\pi n_p(\omega)\left[-\frac{3}{2} \times \frac{1}{3} + \frac{1}{4} \times 2 - \frac{2}{4} \int_{\beta_0}^{+\infty} p(\beta)\, d\beta \right]$$

$$= \frac{\pi}{2}\, n_p(\omega) \int_{\beta_0}^{\infty} \frac{4/\pi}{1 + 4\beta^2}\, d\beta$$

$$= \frac{\pi}{2}\, n_p(\omega)\left[1 - \frac{2}{\pi} \tan^{-1}(2\beta_0) \right] \tag{3.6.17}$$

where β_0 is the normalized distance between the real frequency axis and the pole line. This is the same as the earlier result shown in Eq. (3.6.7).

C. Power and Variance of the Group Delay

We produce power of the group delay pulse process is $\tau = -\Sigma_m \tau_m$ for each strip $d\delta$ (or $d\beta$), where all the pulses have the same shape. If we assume that strips are statistically independent (as we did for the mean), we can add the power from each strip by integrating over β to get the total contribution from all singularities due to both the zero cloud and the pole-line.

The Fourier transform of $\tau_m(\omega)$ (using x as the transform variable) is

$$T_m(x) = \frac{1}{2\pi} \int_{-\infty}^{+\infty} \tau_m(\omega)\, \exp(-j\omega x)\, d\omega$$

$$= (a_m/2)\, \exp(-\delta x) \tag{3.6.18}$$

where

$$\tau_m(\omega) = \frac{a_m \delta}{(\omega - \omega_m^r)^2 + \delta^2} \tag{3.6.19}$$

and $\delta_m = \delta$ on a δ-strip. The power of the pulse train $\tau = -\Sigma_m \tau_m$ is given by

$$P_s(x) = \frac{\langle a_m^2 \rangle_m}{4D_s}\, \exp(-2\delta x)$$

$$= \frac{1}{4D_s}\, \exp(-2\delta x) \tag{3.6.20a}$$

where

$$\langle a_m^2 \rangle_m = 1 \tag{3.6.20b}$$

and D_s is the average pulse separation. That is,

$$D_s = \frac{1}{n_s} \tag{3.6.21}$$

where n_s denotes the density of the singularities.

The density function for the singularities, which is 2-dimensionally distributed in the complex frequency plane, is given by

$$n_s(\beta)\,d\beta = [\tfrac{3}{2}\delta(\beta - \beta_0) + \tfrac{1}{4}p(|\beta|)]n_p\,d\beta,$$

$$\uparrow \qquad\qquad \uparrow$$

$$\text{poles and zeros} \qquad \text{zero}$$
$$\text{on the pole-line} \qquad \text{cloud} \qquad\qquad\qquad (3.6.22)$$

where $\delta(\beta)$ is a delta-function of β. The power due to the singularities on the pole-line is therefore

$$P_{s,\text{pole-line}}(x) = \frac{3n_p}{8} \int_{-\infty}^{+\infty} \exp(-2x\,|\beta|/n_p)\delta(\beta - \beta_0)\,d\beta \qquad (3.6.23)$$

Similarly, the power from the other singularities, i.e. the zero cloud, is expressed as

$$P_{s,\text{zero-cloud}}(x) = \frac{n_p}{4\pi} \int_0^\infty \exp(-2x\beta/n_p)\left(\frac{1}{1 + 4(\beta + \beta_0)^2} + \frac{1}{1 + 4(\beta - \beta_0)^2}\right)d\beta \qquad (3.6.24)$$

Consequently, the total power of the group delay due to all the singularities is written as (see Appendix 1)

$$P_{s,\text{total}}(x) = P_{s,\text{pole-line}}(x) + P_{s,\text{zero-cloud}}(x)$$

$$\approx \tfrac{3}{8} n_p \exp(-2x\delta_0) + \frac{1}{16\pi}\frac{1}{\delta_0^2 x} \qquad (3.6.25)$$

The group delay power has a trend of $1/x$, as $\delta_0 x$ becomes large. We can thus see '$1/f$' type fluctuations in the reverberant phase. And the variance of the group delay decreases in proportion to δ_0^2 and independent of the modal density.

This relationship suggests that the phase variability is mainly due to the zeros (not poles). For high modal overlap, the Cauchy distribution reduces to

$$p(\beta) = \frac{1}{\pi\beta^2} \qquad (3.6.26)$$

and, if we calculate the number of nonminimum phase zeros in a frequency band $\Delta\omega$,

$$N_z^+ = \frac{n_p(\omega)\Delta\omega/4}{M} = \frac{\Delta\omega/4}{\pi\delta_0} \qquad (3.6.27)$$

The number of zeros is independent of the frequency, while the number of poles increases as the frequency increases.

3.6.3. Experimental Results of Reverberation Phase

We have conducted a series of TF measurements in an $86\,\text{m}^3$ room with a $1.8\,\text{s}$ reverberation time in the $500\,\text{Hz}$ octave band (see Fig. 3.6.6). This room has 2700 modes in this octave band, so the maximum possible number of nonminimum phase zeros is 675.

A. *Detecting Zeros in the Complex Domain.* The TF is analysed using impulse response data and FFT methods. Suppose that an impulse response is composed of N sampled data points. Even though such a finite size impulse response does not give eigenvalues of a physical system, the TF phase can be estimated correctly below the real frequency line in the complex domain, when the data record duration exceeds the reverberation time (Tohyama and Lyon, 1989b). The TF for the finite-length impulse response data has $N - 1$ zeros in the complex domain, because the zeros are the roots of an $N - 1$ degree polynomial of z, where $z = \exp(j\omega T)$, T is the sampling period, and ω is the complex frequency. These zeros are distributed inside and outside of the unit circle in the z-plane. The region outside the unit circle corresponds to the lower half-plane in the complex frequency domain. The zeros outside the unit circle are therefore nonminimum phase zeros.

The zeros N_z^+ outside the unit circle of $H(z)$ is given by

$$\frac{-1}{2\pi j} \int_c \frac{H'(z)}{H(z)}\, \mathrm{d}z = N_p - N_z^-$$

$$= N_z^+ \qquad (3.6.28a)$$

since

$$N_p = N_z = N_z^+ + N_z^- = N - 1 \qquad (3.6.28b)$$

where $N_p = N - 1$ and N_z^- shows the number of zeros inside the unit circle (minimum phase zeros). Because the TF is an $N - 1$ degree polynomial of z, all the poles of the polynomial are located at the origin of the z-plane. It is obvious that these 'polynomial poles' are not physical poles (eigenvalues) of the system. The distribution of nonminimum phase zeros can then be estimated by calculating the contour integration along a circle of radius greater than unity (Wang and Itakura, 1989).

The accumulated phase can be calculated by using the above-mentioned contour integral. Equation (3.6.28a) can thus be written as

$$\frac{-1}{2\pi j} \int_c \frac{H'(z)}{H(z)}\, \mathrm{d}z = \frac{-1}{2\pi j} \int_c \frac{\mathrm{d}}{\mathrm{d}z} \ln H(z)\, \mathrm{d}z = N_z^+ \qquad (3.6.29)$$

Since $H(\omega) = |H(\omega)| e^{\,j\phi(\omega)}$, the accumulated phase is given by

$$\phi(\omega) = -2\pi N_z^+ \quad \text{(rad)} \qquad (3.6.30)$$

Here, we can consider that $2N_z^+$ equals the total number of uncancelled poles and nonminimum phase zeros, because the number of uncancelled poles equals the number of nonminimum phase zeros. That is, the calculated phase is the same as the reverberation phase by Eq. (3.6.5). Although the procedure for zero counting is based on finite impulse response data, the accumulated phase can be calculated from the number of nonminimum phase zeros.

We calculated the average number of zeros in four TFs in the reverberant fields for different distances of the microphone from the loudspeaker. The average number of zeros below the test frequency line in the lower half-plane is illustrated in Fig. 3.6.6. The result demonstrates that the distribution of zeros follows the solid line, which decreases inversely proportionally to the distance from the pole-line. The total number of nonminimum phase zeros is 48 at the 500 Hz octave band following Eq. (3.5.17). The distance between the pole-line and the real frequency line is 3.8 ($= \delta_0 = 6.9/T_R$). The computational procedure is described in Appendix 2.

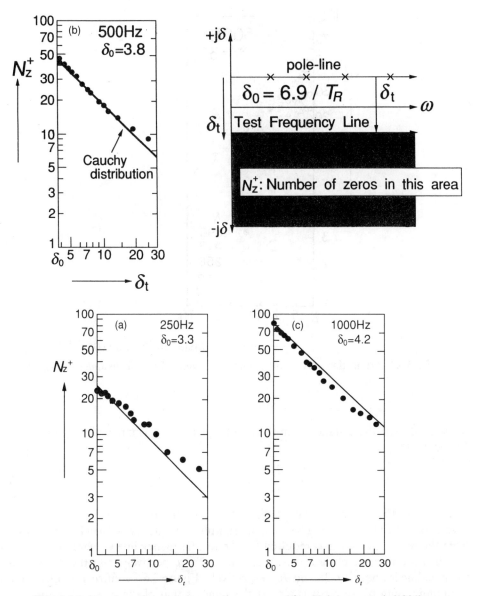

Fig. 3.6.6. Number of nonminimum phase zeros. After Tohyama et al. (1991).

B. *Distribution of Zeros in a Truncated TF.* The distribution of TF zeros in a region far from the pole-line can be estimated by using a truncated finite impulse response record. Figure 3.6.7 shows the distribution of zeros calculated in this manner. We expect that the distribution of zeros follows the true distribution as the distance between the pole line and the real frequency axis becomes large.

Truncation is necessary in most phase analysis applications because of limited data memory capacity and low signal-to-noise ratio in the impulse response data acquisition. Exponential windowing has been proposed as a way of reducing truncation artifacts (Tohyama and Lyon, 1989b). In effect, exponential windowing moves the pole-line farther away from the real frequency axis. There are no severe truncation effects on TF

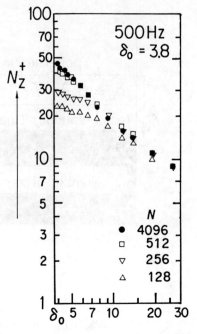

Fig. 3.6.7. Truncation effects on the number of zeros. From Tohyama et al. (1991).

phase analysis if the distance from the real frequency axis is larger than (Tohyama and Lyon, 1989b)

$$\delta = \left(\frac{T_R}{T_L} - 1\right)\delta_0 \qquad (T_L \leqslant T_R) \qquad (3.6.31)$$

where T_L is the truncated data length. For the situation shown in Fig. 3.6.7, the reverberation time is about 1.8 s and the sampling frequency is 1414 Hz. The distance from the real frequency axis required for the test frequency line is therefore 15, 34 and 72 when the number of data is 512, 256 and 128, respectively, and the distance from the pole-line is about 19, 38 and 76, respectively. Figure 3.6.7 confirms that the number of nonminimum phase zeros is almost the same as that of the true distribution in the region far from the pole-line.

We can see, however, that the requirement given by Eq. (3.6.31) is strict. The true zero distribution is observed even in the region near the real frequency axis if the truncation length is longer than $T_R/3$; the distance from the pole-line required to estimate the true distribution of zeros is about 6, 13 and 25, respectively. The effect of truncation is related to both modal separation and damping through the modal overlap parameter. Modal overlap is assumed small in Eq. (3.6.31), but when the modal overlap is large, the rate of change in the frequency response curve is determined by the damping. As Schroeder (1954a, b) has shown, the average distance between peaks in the response curves is $7/T_R$. To define the response curves, the frequency resolution must

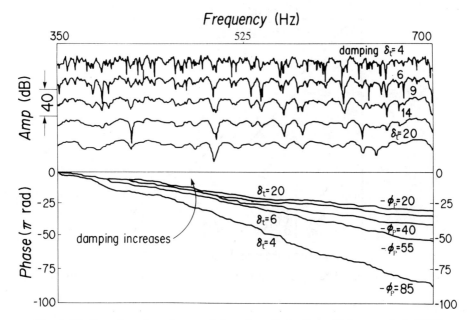

Fig. 3.6.8. Reverberation phase and damping effects. From Tohyama et al. (1994).

be at least $3.5/T_R$. While a longer record length will give slightly better resolution, the data in Fig. 3.6.7 suggest that a shorter record length is acceptable.

C. *Reverberation Phase Samples.* Samples of the amplitude and accumulated phase characteristics of TFs are shown in Fig. 3.6.8. The distance between the pole-line and the test frequency line is changed by applying an exponential window to the impulse response record. The ϕ_p in the graph denotes the theoretical estimation for the accumulated phase from the number of nonminimum phase zeros based on the Cauchy distribution. We can see that the reverberation phase trend is predictable.

D. *Phase Fluctuation Δ-Statistics.* Lyon (1987) pointed out that the phase process can be modelled by a random walk and estimated that the variance of the accumulated phase linearly increases as the frequency increases. Such variances can also be formulated using Δ-statistics (Dyson and Metha, 1963). The idea for the Δ-statistics originally came from nuclear physics (Dyson and Metha, 1963). A plot of $N(E)$, which is the number of levels (energy eigenvalues in a system) having energy between zero and E, looks like a staircase, which gives a good visual impression of the overall regularity of the level series. It is customary to measure the average level spacing by drawing a line having the same average slope as the staircase. Fluctuations in the level series then appear as deviations of the staircase from the straight line.

A suitable method for analysing the fluctuations is to fit a linear function to the level distribution using a least-squares criterion (Fig. 3.6.9). We use this as a measure of the irregularity (fluctuations) in the phase function, although the accumulated phase is not a simple increasing function (Tohyama et al., 1991).

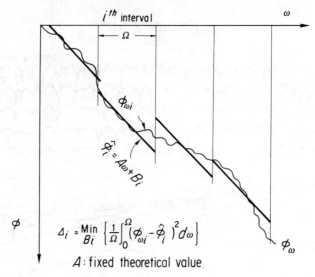

$$\Delta_i = \underset{Bi}{\text{Min}} \left\{ \frac{1}{\Omega} \int_0^\Omega (\phi_{\omega i} - \hat{\phi}_i)^2 d\omega \right\}$$

A : fixed theoretical value

Fig. 3.6.9. Accumulated phase and Δ-statistics. From Tohyama et al. (1994).

The Δ-statistics of the phase is defined by

$$\Delta = \text{Min} \frac{1}{\Omega} \int_0^\Omega [\phi - (-A\omega + B)]^2 \, d\omega \qquad (3.6.32)$$

where Ω is the frequency interval of interest and A is the theoretical trend of the group delay given by $\tau_\infty = 1/2\delta$. The Δ-statistics defined above is estimated for an ensemble of frequency intervals. Figure 3.6.10 illustrates the ensemble average of the Δ-statistics for the reverberation phase. We calculated ensemble-averaged Δ-statistics for four TFs. The calculation procedure is shown in detail in Appendix 3. In Fig. 3.6.10, N denotes the number of frequency points included in a particular frequency interval. The Δ-statistics increase almost linearly when the frequency interval is long, as Lyon (1987) demonstrated using a random walk model; they increase in proportion, however, to the square of the frequency interval when the frequency interval is short.

E. *Group Delay Samples.* Figure 3.6.11 shows samples for the first derivative of the phase. We can see a random pulse train on the frequency axis. The measured TF has finite data records, which means that the TF has only zeros. All the polynomial poles are located at the origin in the discrete z-frequency plane. The positive pulse is due to the minimum phase zeros, and the negative one is due to the nonminimum phase zeros. We can see that the density of the group delay pulses decreases as the distance between the pole-line and the test frequency line increases. This is because the density of the zeros decreases as the distance inceases, following the Cauchy distribution.

The power of the group delay is expressed by the parameter $\delta_0 x$, as shown in Eq. (3.6.25). Figure 3.6.12 shows the power of the group delay at different test frequency lines. We can clearly see a tendency towards the '$1/\delta_t x$' in the group delay following Eq. (3.6.25) as $\delta_t x$ becomes large. Figure 3.6.13 shows the variance in group delay as calculated from the measured TF data. The variances are independent of the frequency band and decrease in proportion to δ_t^2.

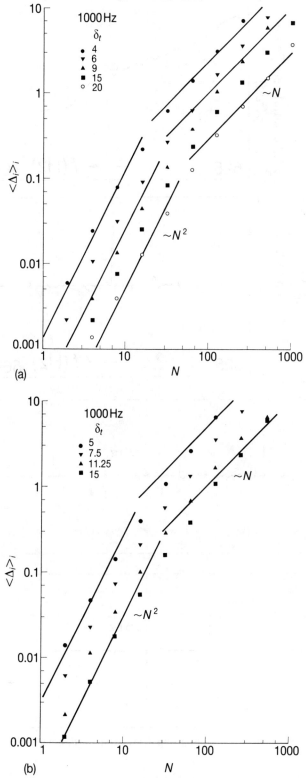

Fig. 3.6.10. Δ-Statistics for the accumulated phase. From Tohyama et al. (1994).

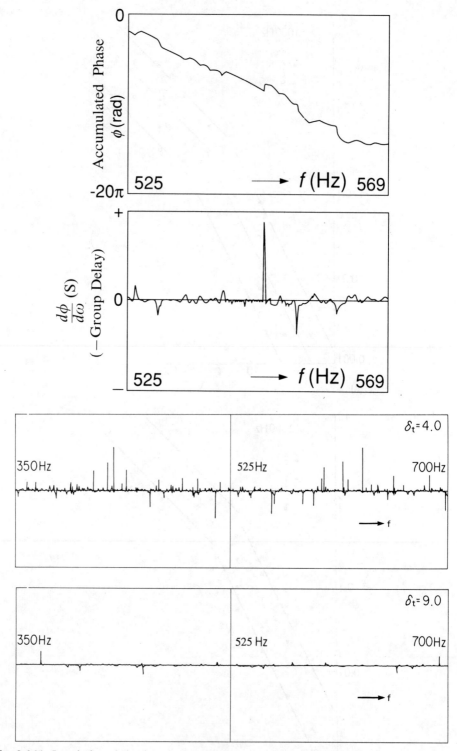

Fig. 3.6.11. Local phase behaviour and group delay samples (first derivative of the phase). After Tohyama et al. (1994).

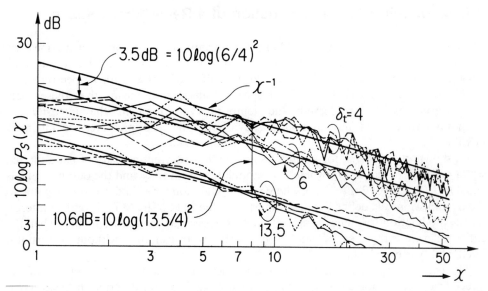

Fig. 3.6.12. Power of group delay samples at different test frequency lines. From Tohyama et al. (1994).

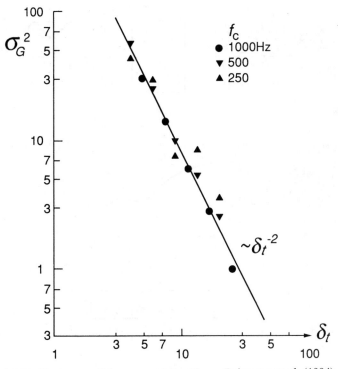

Fig. 3.6.13. Variances of the group delay. From Tohyama et al. (1994).

3.7. Modulation Transfer Function of a Reverberant Space

The modulation transfer function (MTF) was introduced as a room acoustics measure for assessing the effect of an enclosure on speech intelligibility by Houtgast et al. (1980). A speech signal may be regarded as a flow of sound with a spectrum that varies continuously over time. Good speech transmission means that these spectrum variations are preserved; the temporal variations in the intensity, or temporal envelope, at any audiofrequency should be also preserved as much as possible.

As shown in Fig. 3.7.1, for an input signal with a varying intensity of band-filtered white noise $I_i(1 + \cos 2\pi Ft)$, where F = modulation frequency, and the output signal is equal to $I_o(1 + m \cos 2\pi F(t - \tau))$, where m = modulation index and τ = time lag due to the transmission. We define the function $m(F)$ as the MTF. The MTF is only of interest for the modulation frequency range relevant for speech. We may assume that when the entire relevant range, say from 0.4 to 20 Hz, is transmitted without distortions, intelligibility is excellent.

The effects of room reverberation and ambient noise, and the contribution of the direct field, are combined in the single function $m(F)$ as we describe in Section 7.1. When considering only the influence of reverberation, the MTF is the Fourier transform of the squared impulse response (Schroeder, 1981). The input intensity of the

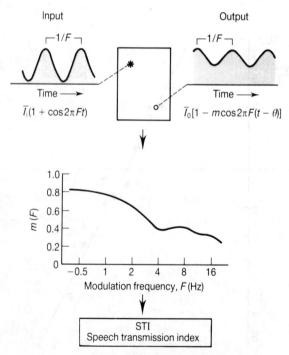

Fig. 3.7.1. Illustration of the MTF concept. From Houtgast et al. (1980).

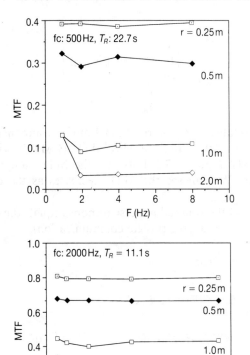

Fig. 3.7.2. Samples of MTF in a large reverberation room. Room volume: $640\,\mathrm{m}^3$, r: distance from the source (m), T_R: reverberation time (s).

band-filtered white noise can be rewritten as

$$I_i(t) = I_i[1 + \mathrm{Re}(e^{j2\pi Ft})] \quad (\mathrm{W/m^2}) \tag{3.7.1}$$

which means the output intensity is given by

$$I_o(t) = \int_0^\infty I_i(t - t')h^2(t')\,dt' = aI_i\left[1 + \frac{1}{a}\,\mathrm{Re}(be^{j2\pi Ft})\right] \quad (\mathrm{W/m^2}) \tag{3.7.2}$$

where

$$a = \int_0^\infty h^2(t')\,dt', \qquad b = \int_0^\infty e^{-j2\pi Ft'}h^2(t')\,dt' \tag{3.7.3}$$

The MTF is therefore

$$m(F) = \frac{\left| \int_0^\infty e^{-j2\pi F t'} h^2(t') \, dt' \right|}{\int_0^\infty h^2(t') \, dt'} \qquad (3.7.4)$$

The MTF equals the modulus of the normalized Fourier component of the squared impulse response. We also define the complex MTF (CMTF) without taking the modulus (Schroeder, 1981). Figure 3.7.2 illustrates the MTF samples measured in a large reverberation room. We can see the MTF decreases as the distance from the sound source increases.

The Fourier transform of the squared impulse response equals the complex autocorrelation function of the TF (Houtgast, private communication),

$$\int_0^\infty e^{-j2\pi F t'} h^2(t') \, dt'$$

$$= \int_{-\infty}^\infty H^*(f') H(f' + F) \, df'$$

$$= \int_{-\infty}^\infty |H^*(f') H(f' + F)| \, e^{-j\{\phi(f' + F) - \phi(f')\}} \, df' \qquad (3.7.5)$$

where H^* denotes the conjugate of the TF $H(f) = |H(f)| \, e^{-j\phi(f)}$. The MTF is thus given by the 'normalized' autocorrelation function of the TF.

4

Visualization of the
Sound Field in a Room

As a sound wave propagates, energy flows through an acoustic medium. Sound radiation mechanisms from loudspeakers or any other types of sound sources are better understood by visualizing the sound field around the source or in a room. The energy flow in the sound field is visualized by use of the sound intensity method (Fahy, 1977, 1989; Chung, 1977). Several types of sound intensities and measurement methods are given in Section 4.1. In Sections 4.2 and 4.3, examples of visualization of sound fields in rooms and around sources are presented. In Section 4.4, sound intensities other than the most commonly used time-average sound intensity are described. The vibration intensity is introduced in this section.

4.1. Sound Intensity and Measurement Method

4.1.1. Instantaneous Sound Intensity

Sound power is well understood by considering analogies between the acoustical quantities and electrical or mechanical quantities. In the impedance analogy, force (pressure times surface area) and particle velocity are analogous to voltage and current, respectively. The sound power transmitted through a unit area (which is termed intensity sound or acoustic intensity) is then given by multiplying the pressure $p(t)$ and the particle velocity $v(t)$ normal to the unit area.

Consider an imaginary surface S normal to the direction of the particle motion, as shown in Fig. 4.1.1. A force acting on the surface is $p(t)S$ assuming that the pressure is constant over the surface. If this surface moves with the velocity $v(t)$, the instantaneous power given to this surface is

$$w(t) = p(t)v(t)S \quad \text{(W)} \tag{4.1.1}$$

Dividing Eq. (4.1.1) by S, the sound power per unit area is obtained as

$$i(t) = p(t)v(t) \quad \text{(W/m}^2\text{)} \tag{4.1.2}$$

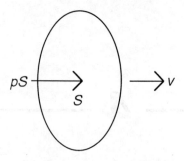

Fig. 4.1.1. Pressure p, particle velocity v, and power on an imaginary surface S.

where $i(t)$ is the instantaneous sound intensity. Since the particle velocity is directional, it is necessary to specify which direction of the particle motion is considered positive. If this reference direction is reversed, the phase of the particle velocity changes by π. If the intensity defined by Eq. (4.1.2) is positive, it is considered that sound energy is transferred in the positive direction in that instantaneous time.

In case of a sinusoidal wave with pressure

$$p(t) = P \cos(2\pi f t) \quad \text{(Pa)} \tag{4.1.3}$$

and particle velocity

$$v(t) = V \cos(2\pi f t + \theta) \quad \text{(m/s)} \tag{4.1.4}$$

$i(t)$ is given by

$$i(t) = \frac{PV}{2} [\cos \theta + \cos(4\pi f t + \theta)] \quad \text{(W/m}^2) \tag{4.1.5}$$

where P and V are amplitudes of the pressure and the particle velocity, respectively, f is the frequency, and θ is the phase difference between them. The instantaneous intensity of the sinusoidal wave is composed of a time-independent term and a time-varying term with twice the wave frequency. The instantaneous intensity itself is very rarely used in real applications because of its time-varying nature.

If $p(t)$ and $v(t)$ are complex, Eq. (4.1.2) should be

$$i_c(t) = p^*(t)v(t) \quad \text{(W/m}^2) \tag{4.1.6}$$

where $i_c(t)$ is the complex sound intensity and * indicates a complex conjugate. If the sinusoids given by Eqs. (4.1.3) and (4.1.4) are expressed, respectively, as

$$p(t) = Pe^{j2\pi f t} \quad \text{(Pa)} \tag{4.1.7}$$

and

$$v(t) = Ve^{j(2\pi f t + \theta)} \quad \text{(m/s)} \tag{4.1.8}$$

the complex sound intensity is given by

$$i_c(t) = PV (\cos \theta + j \sin \theta) \quad \text{(W/m}^2) \tag{4.1.9}$$

which is time-independent. In this case the active intensity is given by

$$i(t) = \tfrac{1}{2} \text{Re}[i_c(t)] \quad \text{(W/m}^2) \tag{4.1.10}$$

A division by 2 is necessary since the powers of the signals given by Eqs. (4.1.7) and (4.1.8) are larger by a factor of 2 than the powers of the corresponding real signals.

4.1.2. Time-average Sound Intensity

In electrical or mechanical engineering, the term 'power' usually means the time-average power, which is used more often than the instantaneous power. The time-average sound intensity is defined as

$$I = \lim_{T \to \infty} \frac{1}{T} \int_{-T/2}^{T/2} i(t) \, dt \quad (\text{W/m}^2) \tag{4.1.11}$$

For a sinusoidal wave, I is given by the first term of Eq. (4.1.5) as

$$I = \tfrac{1}{2} PV (\cos \theta) \quad (\text{W/m}^2) \tag{4.1.12}$$

Figure 4.1.2 shows the pressure (solid line), particle velocity (chain line) and instantaneous intensity (dashed line) for $\theta = 0$, $-\pi/3$, and $-2\pi/3$. It should be noted that the frequency of the instantaneous intensity (a.c. component) is twice the frequency of the pressure and the particle velocity. The time-average intensity (thick solid line) is not a function of time but its levels are shown for convenience. The time-average sound intensity becomes negative if θ is in the second or third quadrant, indicating that the energy flows in the opposite direction. The time-average intensity is commonly used because it is independent of time and, therefore, it is convenient in graphically representing the sound energy flow.

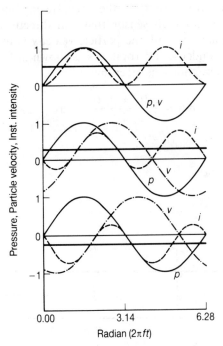

Fig. 4.1.2. Pressure, particle velocity and instantaneous intensity of a sinusoid for various phases.

Equation (4.1.11) is equal to $\Phi_{pv}(\tau)$ with $\tau = 0$, the cross correlation between $p(t)$ and $v(t)$ for stationary signals (see Eq. (2.2.14)). Then the time-average sound intensity is given by the inverse Fourier transform of $S_{pv}(f)$, the cross spectrum between $p(t)$ and $v(t)$, as

$$I = \int_{-\infty}^{\infty} S_{pv}(f) \, df \quad (\text{W/m}^2) \tag{4.1.13}$$

The complex amplitudes of Eqs. (4.1.7) and (4.1.8) are given, respectively, by

$$P(f) = P \quad (\text{Pa/Hz}) \tag{4.1.14}$$

and

$$V(f) = V e^{j\theta} \quad (\text{m/s} \cdot \text{Hz}) \tag{4.1.15}$$

Then the time-average intensity spectrum is given by

$$I(f) = \tfrac{1}{2}\text{Re}[P^*(f)V(f)] = \tfrac{1}{2}PV(\cos\theta) \quad (\text{W/m}^2 \cdot \text{Hz}) \tag{4.1.16}$$

which corresponds to Eq. (4.1.12).

4.1.3. Measurement Method

A. *Two-microphone Method*

The sound intensity is determined by the pressure and the particle velocity. It is therefore necessary to measure those two physical quantities directly or indirectly. A condenser microphone is mostly used for the accurate measurement of the sound pressure. However, there is no such sensor that can directly measure the particle velocity with a high accuracy. Instead, the particle velocity is indirectly measured by use of two closely located condenser microphones as shown in Fig. 4.1.3. The pressure at the centre of the two microphones is given by

$$p(t) = \frac{p_1(t) + p_2(t)}{2} \quad (\text{Pa}) \tag{4.1.17}$$

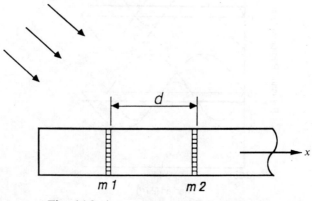

Fig. 4.1.3. A p-p type sound intensity probe.

or in the frequency domain,

$$P(f) = \frac{P_1(f) + P_2(f)}{2} \quad \text{(Pa/Hz)} \tag{4.1.18}$$

where $p_1(t)$ and $p_2(t)$ are the outputs of microphones 1 and 2, respectively, and $P(f)$ is the Fourier transform of $p(t)$. The relation between the particle velocity in the x-direction, $v_x(t)$, and the sound pressure $p(t)$ is

$$v_x(t) = -\frac{1}{\rho} \int_{-\infty}^{t} \frac{\partial p(t')}{\partial x} \, dt' \quad \text{(m/s)} \tag{4.1.19}$$

where ρ is the air density. By the first-order approximation, the above equation is rewritten as

$$v_x(t) = -\frac{1}{\rho d} \int_{-\infty}^{t} [p_2(t') - p_1(t')] \, dt' \quad \text{(m/s)} \tag{4.1.20}$$

where d is the microphone separation. In the frequency domain, it is expressed as

$$V_x(f) = -\frac{P_2(f) - P_1(f)}{j\rho ckd} \quad \text{(m/s·Hz)} \tag{4.1.21}$$

where c is the sound velocity and k is the wavenumber. The sound intensity measurement method using Eqs. (4.1.17) and (4.1.20) or Eqs. (4.1.18) and (4.1.21) is referred to as the two-microphone method. A specially designed probe with two microphones as in Fig. 4.1.3 is called a p-p type intensity probe.

A vector representation of $P_1(f)$, $P_2(f)$ and $V_x(f)$ is shown in Fig. 4.1.4. When the sound propagates in the positive x-direction, $P_1(f)$ is ahead of $P_2(f)$ in phase. $[P_1(f) - P_2(f)]$ is shown by the dotted line and $V_x(f)$ is obtained by rotating $[P_1(f) - P_2(f)]$ by $-\pi/2$ and then dividing by ρckd. If the magnitudes and phases of $P_1(f)$ and $P_2(f)$ are approximately the same, the magnitude of $V_x(f)$ is proportional to the phase difference between $P_1(f)$ and $P_2(f)$, and its phase is approximately the same as those of $P_1(f)$ and $P_2(f)$, resulting in a small phase difference between $P(f)$ and $V_x(f)$ (θ in Eq. (4.1.4)). If the phase difference between $P_1(f)$ and $P_2(f)$ is very

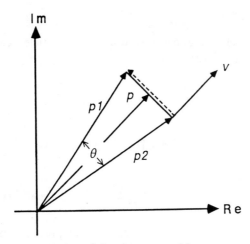

Fig. 4.1.4. A vector representation of signals measured by a p-p type intensity probe.

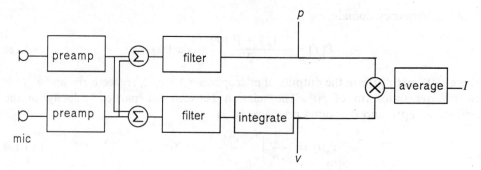

Fig. 4.1.5. Block diagram of the direct integration method.

small and their magnitudes are significantly different, the phase difference between $P(f)$ and $V_x(f)$ is close to $\pm\pi/2$, resulting in a small intensity. This type of sound field is called 'reactive'.

B. *Direct Integration Method*

By applying Eqs. (4.1.17) and (4.1.20) to Eq. (4.1.11), the intensity averaged over the range $[t - T, t]$, is obtained as

$$I = \frac{1}{2\rho dT} \int_{t-T}^{t} \left\{ \left([p_1(t') + p_2(t')] \int_{-\infty}^{t'} [p_1(t'') - p_2(t'')] \, dt'' \right) dt' \right\} \quad \text{(W/m}^2) \qquad (4.1.22)$$

The suffix x indicating the direction of measurement has been omitted for simplicity. If a band-limited intensity is required, $p_1(t')$ and $p_2(t')$ are filtered and then Eq. (4.1.22) is applied. This method is called the direct integration method and its block diagram is shown in Fig. 4.1.5. The averaging time T can be varied depending on the purpose of the measurement. One disadvantage of this technique is that a large number of filters is necessary if different bands must be measured simultaneously.

C. *Cross Spectrum Method*

The cross spectrum between the pressure and the particle velocity gives the sound intensity spectrum (Eq. 2.2.15).

$$S_{pv}(f) = \lim_{T \to \infty} E[P^*(f)V(f)/T]$$

$$= -\frac{1}{4j\pi f\rho d} \lim_{T \to \infty} E[\{P_2^*(f) + P_1^*(f)\}\{P_2(f) - P_1(f)\}/T] \qquad (4.1.23)$$

$$= -\frac{1}{2\pi f\rho d} \text{Im}[S_{p1p2}(f)] + j \frac{1}{4\pi f\rho d} [S_{p2p2}(f) - S_{p1p1}(f)] \quad \text{(W/m}^2 \cdot \text{Hz)}$$

The intensity spectrum at frequency f consists of positive and negative frequency components. Since $p_1(t)$ and $p_2(t)$ are real, the imaginary part of the cross spectrum

$\mathrm{Im}[S_{p1p2}(f)]$ is an odd function and, therefore, the first term of Eq. (4.1.23) is an even function. The second term (imaginary part) of Eq. (4.1.23) is an odd function. Then the intensity spectrum $I(f)$ is given by

$$I(f) = -\frac{\mathrm{Im}[G_{p1p2}(f)]}{2\pi f\rho d} \quad (\mathrm{W/m^2 \cdot Hz}) \tag{4.1.24}$$

where $G_{p1p2}(f)$ is the one-sided cross spectrum between $p_1(t)$ and $p_2(t)$. The sound intensity measurement method using Eq. (4.1.24) is called a cross spectrum method. This method is suitable for a two-channel FFT (fast Fourier transform) analyser since the cross spectrum is very easily obtained by the analyser. The largest advantage of this method is that the frequency resolution is much finer than that of the filters used in the direct integration method. A disadvantage is that the highest frequency of the real-time measurement is limited by the processing speed of the FFT analyser or a specially designed processor.

4.2. Intensity Field in a Room

4.2.1. Propagating Wave

A. *One-dimensional Propagating Wave*

The pressure of a plane propagating wave in the positive x-direction is given by

$$p(x, t) = P\cos(2\pi ft - kx) \quad (\mathrm{Pa}) \tag{4.2.1}$$

where the particle velocity is given by use of Eq. (4.1.19) such as

$$v(x, t) = \frac{P}{\rho c}\cos(2\pi ft - kx) \quad (\mathrm{m/s}) \tag{4.2.2}$$

Then the time-average intensity, which is constant everywhere, is given by

$$I = \frac{P^2}{2\rho c} \quad (\mathrm{W/m^2}) \tag{4.2.3}$$

This is the time-average power transmitted in the positive x-direction through a unit area. The 1-dimensional propagating wave is fairly easily realized in an acoustic tube.

B. *Two-dimensional Propagating Wave Near a Reflecting Wall*

When a plane propagating wave is reflected by an infinite wall as shown in Fig. 4.2.1, the pressure sum of the incident and reflected waves is given by

$$p(x, y, t) = P[\cos(2\pi ft - kx\sin\theta + ky\cos\theta) + \cos(2\pi ft - kx\sin\theta - ky\cos\theta)]$$

$$= P\cos(2\pi ft - kx\sin\theta)[2\cos(ky\cos\theta)] \quad (\mathrm{Pa}) \tag{4.2.4}$$

The pressure amplitude is given by

$$|p(x, y, t)| = |2P\cos(ky\cos\theta)| \quad (\mathrm{Pa}) \tag{4.2.5}$$

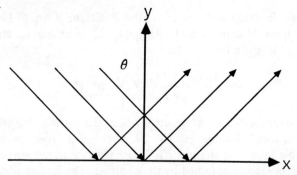

Fig. 4.2.1. Incident and reflected waves by an infinitely large baffle.

The particle velocities in the x- and y-directions are given, respectively, by (see Eq. (4.1.19))

$$v_x(x, y, t) = \frac{2P}{\rho c} \cos(2\pi f t - kx \sin \theta) \sin \theta \cos(ky \cos \theta) \quad \text{(m/s)} \tag{4.2.6}$$

and

$$v_y(x, y, t) = \frac{2P}{\rho c} \sin(2\pi f t - kx \sin \theta) \cos \theta \sin(ky \cos \theta) \quad \text{(m/s)} \tag{4.2.7}$$

The locus of the particle velocity is elliptical with

$$\frac{|v_x(x, y, t)|^2}{a^2} + \frac{|v_y(x, y, t)|^2}{b^2} = 1 \tag{4.2.8}$$

where

$$a = \frac{2P}{\rho c} \sin \theta \cos(ky \cos \theta) \tag{4.2.9}$$

and

$$b = \frac{2P}{\rho c} \cos \theta \sin(ky \cos \theta) \tag{4.2.10}$$

The intensities in the x- and y-directions are given, respectively, by

$$I_x = 2 \frac{P^2}{\rho c} \sin \theta \cos^2(ky \cos \theta) \quad \text{(W/m}^2\text{)} \tag{4.2.11}$$

and

$$I_y = 0 \quad \text{(W/m}^2\text{)} \tag{4.2.12}$$

The intensity in the x-direction is dependent only on y. The intensity in the y-direction is always zero.

The pressure amplitude, particle velocity locus, and time-average sound intensity are shown in Fig. 4.2.2. The pressure and the intensity are largest on the wall surface. At

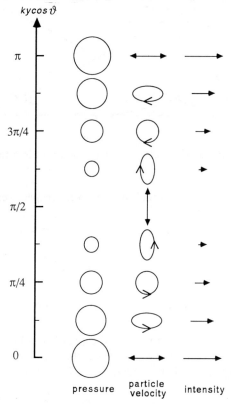

Fig. 4.2.2. Pressure amplitude, particle velocity locus and time-average sound intensity of a plane propagating wave near an infinitely large baffle.

the normalized distance from the wall, $ky \cos \theta = (2n - 1)\pi/2$ $(n = 1, 2, 3, \ldots)$, the pressure and the intensity are equal to zero and the particle motion is only in the vertical direction. The intensity is always in the positive x-direction even though the direction of the particle motion is reversed at every $\pi/2$ period.

4.2.2. Standing Wave

A. One-dimensional Standing Wave

One-dimensional standing waves can exist in an acoustic tube with two rigid ends. A pressure distribution inside the tube that satisfies the boundary condition is given as

$$p(x, t) = P_n \cos(n\pi x/L) \cos(2\pi f_n t) \quad (0 \leqslant x \leqslant L) \tag{4.2.13}$$

where n is the mode number, $f_n (= nc/2L)$ is the resonance frequency, and L is the length of the tube. Note that the spatial term is separated from the time-dependent term. This is true for any types of standing waves. A standing wave has the same phase at any location in the space (except 180° difference). The particle velocity and the

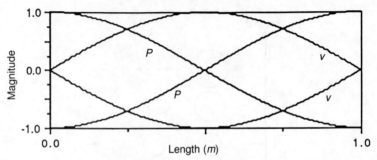

Fig. 4.2.3. Pressure and particle velocity distributions in a tube with rigid ends.

instantaneous intensity in the x-direction are given, respectively, by

$$v_x(x, t) = \frac{P_n}{\rho c} \sin(n\pi x/L) \sin(2\pi f_n t) \tag{4.2.14}$$

and

$$i_x(x, t) = \frac{P_n^2}{4\rho c} \sin(2n\pi x/L) \sin(4\pi f_n t) \tag{4.2.15}$$

Equations (4.2.13) and (4.2.14) show that the pressure and the particle velocity have 90° phase difference everywhere and, therefore, there is no time-average intensity flow in the 1-dimensional tube with two rigid ends. Figure 4.2.3 shows the pressure and particle velocity distributions for $n = 1$.

B. *Two-dimensional Standing Wave*

The pressure distribution of the (m, n)th mode in a 2-dimensional rectangular space of size (L_x, L_y) is given by

$$p(x, y, t) = P_{mn} \cos(m\pi x/L_x) \cos(n\pi y/L_y) \cos(2\pi f_{mn} t) \tag{4.2.16}$$

where m and n are equal to the number of nodal lines in the x- and y-directions, respectively. P_{mn} and f_{mn} are the amplitude and resonance frequency of the (m, n)th mode, respectively. The particle velocities in the x- and y-directions are given by,

$$v_x(x, y, t) = P_{mn} \frac{m\pi}{\rho c k_{mn} L_x} \sin(m\pi x/L_x) \cos(n\pi y/L_y) \sin(2\pi f_{mn} t) \tag{4.2.17}$$

and

$$v_y(x, y, t) = P_{mn} \frac{n\pi}{\rho c k_{mn} L_y} \cos(m\pi x/L_x) \sin(n\pi y/L_y) \sin(2\pi f_{mn} t) \tag{4.2.18}$$

Since the pressure and the particle velocities are 90° out of phase with each other, there is no time-average sound intensity in this case. This is always true if there exists only a single mode in the space.

If there are more than one mode in the 2- or 3-dimensional space with some phase difference, however, the time-average sound intensity can exist (Suzuki, 1981b). Let us

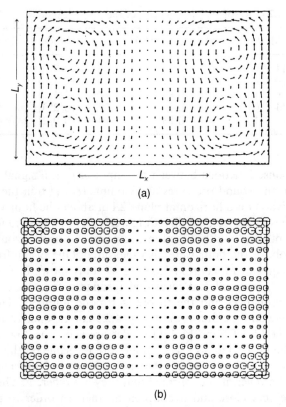

(a)

(b)

Fig. 4.2.4. Calculated intensity vortex (a) and pressure distribution (b) in a 2-dimensional space.

consider a pressure distribution with a $(3, 1)$ mode and a $(1, 2)$ mode with the same amplitude and with the 90° phase difference. If L_y/L_x is $(3/8)^{1/2}$, these two modes have the same wavenumber (same resonance frequency). The pressure is given as a sum of the terms obtained by replacing (m, n) by $(3, 1)$ and $(1, 2)$. The particle velocities in the x- and y-directions are obtained by the same way. The calculated pressure and intensity vectors are shown in Fig. 4.2.4. The energy can only circulate since there are no sources and sinks in the space for a pure standing wave. This was confirmed by intensity measurement in a 3-dimensional reverberation room (see Section 4.2.3).

4.2.3. Wave Field in a Reverberation Room

A. *Modal Analysis of Standing Waves*

In this section, experimental studies of the sound field in a 3-dimensional reverberation room are described. The modal analysis method, which is normally used for the vibration analysis of mechanical structures, is applied to analyse the sound field.

A reverberation room used for the experiment has the dimensions 8.4 m(d) × 6.7 m(w) × 5.3 m(h) with parallel walls. Theoretically calculated resonance frequencies of the first ten modes are shown in Table 4.2.1.

Table 4.2.1. Resonance frequencies of the first ten modes.

Mode	Frequency (Hz)		Mode	Frequency
1	$[1,0,0]$	20.19	6	$[2,0,0]$	40.37
2	$[0,1,0]$	25.31	7	$[0,1,1]$	40.79
3	$[0,0,1]$	31.99	8	$[1,1,1]$	45.51
4	$[1,1,0]$	32.37	9	$[0,2,0]$	50.62
5	$[1,0,1]$	37.83	10	$[0,0,2]$	63.99

Frequency response functions between the input electrical signal to a loudspeaker placed at a corner and sound pressures at 48 points (8 and 6 in the depth and width directions, respectively) on a horizontal plane 2.1 m above the floor were measured. In order to see changes in the vertical direction, frequency response functions at 8 points (2 and 4 in the depth and width directions, respectively) on horizontal planes 2.76 m, 3.42 m and 4.1 m above the floor were also measured. One of the frequency response functions in the frequency range from 22 Hz to 47 Hz is shown in Fig. 4.2.5. From the comparison of the theoretical and measured resonance frequencies it is possible to match measured resonances and the theoretical mode numbers. The lowest $[1,0,0]$ mode is out of range and the first peak in Fig. 4.2.5 corresponds to the $[0,1,0]$ mode. Modal shapes of the $[1,1,0]$ and $[1,1,1]$ modes are shown in Fig. 4.2.6a and b. The modal shapes of the $[1,1,0]$ and $[1,1,1]$ modes are the same on the lowest horizontal plane (2.1 m above the floor). From the measurement of the modes on different horizontal planes, however, it is confirmed that they are different. The $[2,0,0]$ and the $[0,1,1]$ modes are very close and overlap each other. In order to properly separate these overlapped modes, the polyreference curve-fitting technique must be used (for example, Richardson and Formenti, 1985).

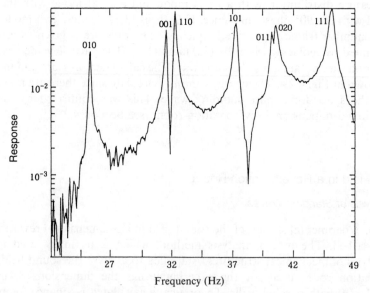

Fig. 4.2.5. An example of frequency response functions.

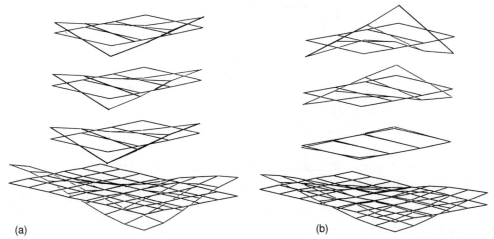

(a) (b)

Fig. 4.2.6. Modal shapes of (a) the $[1, 1, 0]$ and (b) the $[1, 1, 1]$ modes in a reverberation room.

By applying the modal analysis to the 3-dimensional sound field, it becomes possible to identify each mode not only by the frequency matching but also by the mode shape.

B. *Energy Circulation in a Reverberation Room*

In a reverberation room there is a very small energy absorption and, therefore, there is a very small time-average energy flow. If the sound intensity method is applied in the reverberation room, it is almost impossible to measure the intensity with sufficient accuracy, because the phase difference between pressures at two points is very small compared to the phase mismatch between pressures at the two channels of a measuring instrument. In special conditions, however, it is possible to generate an efficient sound energy flow in the reverberation room (see Section 4.2.2.B).

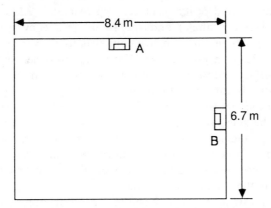

Fig. 4.2.7. A reverberation room with two loudspeakers A and B.

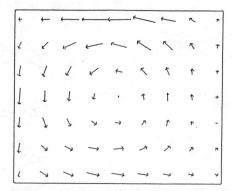

Fig. 4.2.8. Measured energy circulation in a reverberation room.

For the sound energy to flow in a 2- or 3-dimensional space, at least two modes with some phase difference must be generated at a single frequency. In the same reverberation room introduced above, the $[1,0,0]$ and $[0,1,0]$ modes were generated by two loudspeakers placed along the two adjacent side walls as shown in Fig. 4.2.7. The drive frequency was chosen at 22.5 Hz in order to generate two modes with approximately the same amplitudes. The phase difference between the two signals that were applied to the two loudspeakers was finely adjusted so that one mode leads the other by 90°. Measured intensity vectors are shown in Fig. 4.2.8. The circulation of sound energy is clearly shown by the figure. The reason the sound energy flow can exist in the reverberation room is that the pressure of one mode and the particle velocity of the other mode are in-phase if the two modes are excited with a 90° phase difference.

4.2.4. Intensity Near Absorbing Materials

The sound energy absorption phenomenon by an acoustic material is visualized by use of the sound intensity method. Figure 4.2.9 shows the pressure distribution (top), intensity vectors (middle), and surface acoustic impedance (bottom) of polyurethane foams (50 mm and 150 mm $\times$ 400 mm $\times$ 400 mm) placed on a rigid floor in a semi-anechoic chamber. A loudspeaker is placed approximately 1.5 m above the centre of the absorber (zero incident angle) and the signal frequency is 255 Hz. Instead of a two-microphone probe, a single microphone was used to measure pressure signals at two points ($d = 42$ mm in Eq. (4.1.24)). For a thin absorber (left), the main absorbing area is the top surface and the maximum pressure is also on the top surface. The specific acoustic impedance is almost uniform all over the top surface. (This is not always true for other incident angles.) For a thick foam (right), a large portion of sound energy is absorbed from the side surfaces. The acoustic impedance is very uniform in this case but the value is quite different from that of the thin foam. The phase difference between the pressure and the velocity (dotted line in the bottom chart) near the surface is close to $\pm 90°$, indicating that the sound field is reactive even near the absorbing material.

It is clear that the visualization of the sound field by use of the pressure and the intensity distributions helps us understand the absorption mechanism of a material andeffects of the material geometry. An acoustic material behaves differently at different

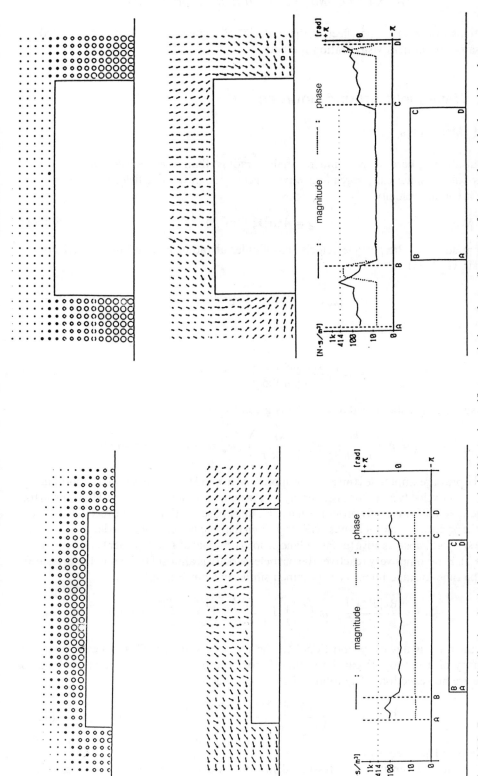

Fig. 4.2.9. Pressure distribution (top), intensity vectors (middle), and specific acoustic impedance (bottom) of polyurethane blocks with volumes of $0.4 \times 0.4 \times 0.05\,\mathrm{m}^3$ (left) and $0.4 \times 0.4 \times 0.15\,\mathrm{m}^3$ (right).

locations in a room. The effect of the material placement can also be investigated by the same sound field visualization method.

4.3. Wave Field Around Sources

4.3.1. Monopoles

A pulsating sphere with a much smaller radius than the wavelength is called a monopole. A spherically spreading wave is generated in the free field. For a monopole with a volume velocity

$$Q = 4\pi a^2 V_a \quad (\text{m}^3/\text{s}) \tag{4.3.1}$$

the pressure and the particle velocity amplitudes at radius r are given, respectively, by (Skudrzyk, 1971)

$$P = j\rho c V_a \frac{jka}{1 + jka}\left(\frac{ka}{kr}\right) e^{-jk(r-a)} \quad (\text{Pa}) \tag{4.3.2}$$

and

$$V = V_a \frac{1 + jkr}{1 + jka}\left(\frac{ka}{kr}\right)^2 e^{-jk(r-a)} \quad (\text{m/s}) \tag{4.3.3}$$

The specific acoustic impedance (P/V) is given by

$$z_a = \rho c\left(\frac{(kr)^2}{1 + k^2 r^2} + j\frac{kr}{1 + k^2 r^2}\right) = r_a(r) + jx_a(r) \quad (\text{N s/m}^3) \tag{4.3.4}$$

The pressure amplitude is inversely proportional to r and the phase delay is proportional to $(r - a)$. The velocity amplitude is inversely proportional to r for large $kr\,(\gg 1)$. However, in the range $kr \ll 1$, it is inversely proportional to r^2. The phase relationship between the pressure and the particle velocity at radius r is known from the specific acoustic impedance. For a very large kr, z_a is almost resistive (as for a plane propagating wave). On the other hand, for a small $kr\,(\ll 1)$, z_a becomes very reactive (the particle velocity is almost 90° behind the pressure). The time-average intensity is obtained similarly from Eq. (4.1.16) as

$$I = \frac{\text{Re}[P^*V]}{2} = \rho c\left(\frac{ka}{kr}\right)^2 \left(\frac{(ka)^2}{1 + k^2 a^2}\right)\frac{V_a^2}{2} = \frac{|P|^2}{2\rho c} \quad (\text{W/m}^2) \tag{4.3.5}$$

The intensity is always proportional to the pressure squared. This indicates that the intensity of a monopole can be obtained from the pressure measurement as long as there are no reflections. The total radiated power is

$$W = 4\pi a^2 \rho c\left(\frac{(ka)^2}{1 + k^2 a^2}\right)\frac{V_a^2}{2} = \frac{4\pi a^2 r_a(a)V_a^2}{2} \quad (\text{W}) \tag{4.3.6}$$

which is independent of r.

The sound intensity and pressure distributions around monopoles are easily calculated. The intensity field around two monopoles with the normalized distance

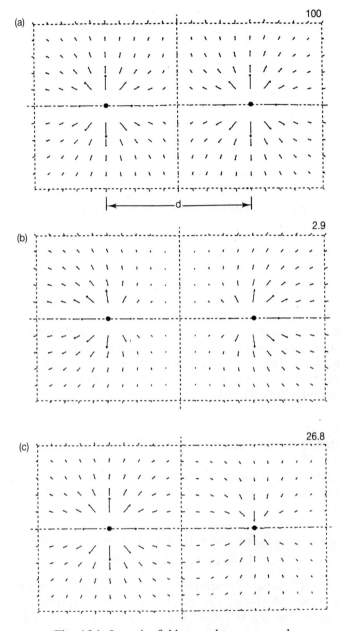

Fig. 4.3.1. Intensity field around two monopoles.

$kd = 0.5$ are shown by Fig. 4.3.1a, b, and c for three cases with $(1,1)$, $(1,-1)$ and $(1,-0.5)$, respectively. The first and the second numbers in the parentheses represent the respective volume velocities of the two monopoles, and a negative sign indicates that the monopoles vibrate 180° out of phase with each other. The two monopoles for the first two cases radiate the same amount of energy since they are balanced and neither of them leads in relative phase. However, the magnitude of the maximum intensity vector for the $(1, -1)$ case is approximately 3/100 of the $(1, 1)$ case (relative

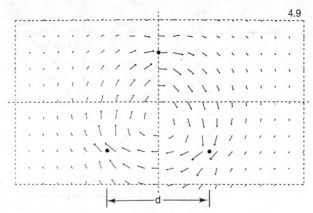

4.9

Fig. 4.3.2. Intensity field around three monopoles.

magnitude for each case is shown at the top right corner of each chart). If the right source is driven out-of-phase with one half of the volume velocity of the left source, it absorbs energy as shown by Fig. 4.3.1c.

The phase relationship between coherent sources is very important in radiation and absorption of sound. Figure 4.3.2 shows the intensity field around three monopoles located at apexes of a regular triangle. The sources are separated by distance d and they are vibrating with 120° phase differences with each other ($kd = 0.5$). The phase difference is positive in the counterclockwise direction. The energy circulates in the clockwise direction. This is due to the property of radiation that a source that leads in phase radiates sound more effectively.

4.3.2. Sound Radiation and Absorption by Two Sources

Let us assume that sources 1 and 2 are vibrating with velocities V_1 and V_2, respectively, in a free or closed space. Forces F_1 and F_2 acting on sources 1 and 2, respectively, are given by

$$F_1 = Z_{11}V_1 + Z_{12}V_2 \quad (N)$$

$$F_2 = Z_{21}V_1 + Z_{22}V_2 \quad (N)$$

(4.3.7)

where Z_{11} and Z_{22} (N s/m) are the self radiation (mechanical) impedance of sources 1 and 2, respectively, and $Z_{12}(=Z_{21})$ is the mutual radiation (mechanical) impedance. Notice that radiation impedances include the effect of the boundary reflections.

The radiated power from source 1 is given by

$$W_1 = \frac{\text{Re}[F_1^*V_1]}{2}$$

$$= \frac{R_{11}|V_1|^2}{2} + \frac{|V_1 V_2|}{2}(R_{12}\cos\theta + X_{12}\sin\theta) \quad (W)$$

(4.3.8)

where R_{11} is the self radiation resistance of source 1, R_{12} and X_{12} are the real and imaginary parts of Z_{12}, and θ is the relative phase lead of source 1 to source 2. The radiated power from source 2 is given by a similar equation. Then the total radiated power is given by

$$W_t = \frac{R_{11}|V_1|^2}{2} + \frac{R_{22}|V_2|^2}{2} + R_{12}|V_1 V_2| \cos\theta \quad \text{(W)} \tag{4.3.9}$$

where R_{22} is the self radiation resistance of source 2. Since W_t is always positive, the following relationship must be satisfied:

$$R_{12}^2 < R_{11}R_{22} \tag{4.3.10}$$

The power radiated from one of the sources can be negative (energy sink) if the mutual radiation effect is strong enough to cancel the self radiation. As seen from Eq. (4.3.8), if the two sources are vibrating antiphase ($\theta = \pm\pi$), the radiated power from source 1 is negative if

$$R_{11}|V_1| < R_{12}|V_2| \tag{4.3.11}$$

Since self and mutual radiation impedances of two spherical sources with a distance d and the same radius a are given, respectively, by

$$Z_{11} = 4\pi a^2 \rho c k^2 a^2 \left(\frac{1+j}{ka}\right) \quad \text{(N·s/m)} \tag{4.3.12}$$

and

$$Z_{12} = 4\pi a^2 \rho c k^2 a^2 \left(\frac{(\sin kd)}{kd} + \frac{j(\cos kd)}{kd}\right) \quad \text{(N·s/m)} \tag{4.3.13}$$

The real and imaginary parts of the mutual radiation impedance are proportional to $(\sin kd)/kd$ and $(\cos kd)/kd$, respectively, which are shown in Fig. 4.3.3. Source 1 can be a sound energy sink if

$$\left|\frac{V_1}{V_2}\right| < \frac{(\sin kd)}{kd} \tag{4.3.14}$$

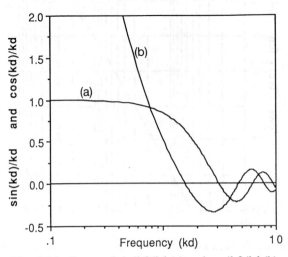

Fig. 4.3.3. Curves of $\sin(kd)/kd$ (a) and $\cos(kd)/kd$ (b).

This condition is rather easily satisfied in the low-frequency region $(kd < 1)$ if V_1 is smaller than V_2.

On the other hand, if the phase lead of source 1 is $-\pi/2$, then the radiated power from source 1 is negative if

$$\left|\frac{V_1}{V_2}\right| < \frac{(\cos kd)}{kd} \tag{4.3.15}$$

Since $(\cos kd)/kd$ can take a very large value for a very small $kd\,(\ll 1)$, source 1 may absorb sound energy even if it vibrates with a larger amplitude than source 2. A source leading in phase tends to radiate more energy than a source vibrating with a phase lag as long as X_{12} is positive. The total radiated power remains the same as that radiated by the two sources without mutual interaction if $\theta = \pm\pi/2$.

In active noise control, a secondary source is added to reduce the total radiated power or to reduce pressures at specified points. The former case is discussed here. Assuming that V_1 is fixed, conditions to minimize W_t can be found by differentiating W_t with respect to $|V_2|$ and θ. These are

$$\theta = \pi\ (R_{12} > 0) \qquad \text{or} \qquad 0\ (R_{12} < 0) \tag{4.3.16}$$

and

$$|V_2| = |V_1||R_{12}/R_{22}| \tag{4.3.17}$$

The minimized total radiated power is given by

$$W_m = \frac{R_{11}|V_1|^2}{2}\left(1 - \frac{R_{12}^2}{R_{11}R_{22}}\right) \tag{4.3.18}$$

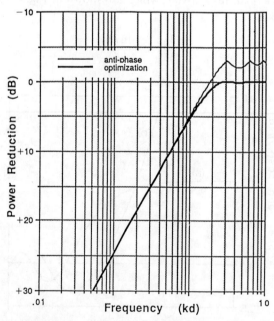

Fig. 4.3.4. Power reduction when two monopoles are driven antiphase (same volume velocity) and when they are driven by the optimization condition.

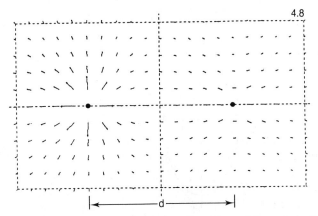

Fig. 4.3.5. Intensity fields around two monopoles vibrating antiphase and with $V_2 = -0.959V_1$ (optimization).

which is equal to W_1, and source 2 is not radiating energy at all but only working to reduce the radiated power from source 1.

$$\eta = 1 - \frac{R_{12}^2}{R_{11}R_{22}} \qquad (4.3.19)$$

is a coefficient that is an indicator of radiated power reduction.

Figure 4.3.4 shows the reduction of total radiated power when source 2 is added to minimize the total radiated power while source 1 is vibrating with a fixed volume velocity (optimization). The figure also shows the power reduction by the antiphase drive. The two cases give almost the same power reduction below $kd = 1$. Therefore, if the power reduction is more than 5 dB, the effectiveness of the two cases is considered the same. In the range $kd \geqslant 3$, the antiphase case gives an increase of radiated power by 3 dB since there is no effective interference between the two sources.

Figure 4.3.5 shows intensity patterns around the two monopoles for the optimization condition ($kd = 0.5$ and $V_2 = -0.959V_1$). Compared to the antiphase case (Fig. 4.3.1b), the intensity field is quite different even though the power reductions are about the same (see Fig. 4.3.4). The figure clearly shows that the additional source (right side) radiates no energy in the optimized condition.

4.3.3. Flat Piston in an Infinite Baffle

The solution for sound radiation from a circular flat piston in an infinite baffle as shown in Fig. 4.3.6 is well known (Kinsler and Frey, 1962; Skudrzyk, 1971). The pressure amplitude at point (x, y, z) is given by the Huygens–Rayleigh integral as

$$P(x, y, z) = j\omega\rho \int_S \frac{V(x', y', 0)}{2\pi r'} e^{-jkr'} dx' dy' \quad \text{(Pa)} \qquad (4.3.20)$$

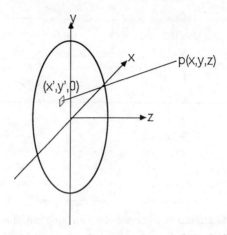

Fig. 4.3.6. A circular flat piston in an infinite baffle.

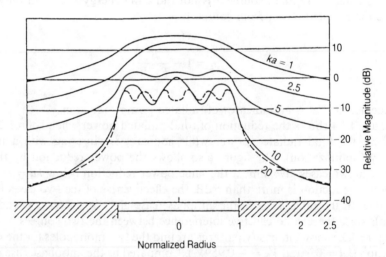

Fig. 4.3.7. Pressure distributions in front of a circular flat piston with radius a in an infinite baffle. From (Kido 1988).

where $V(x', y', 0)$ is equal to V_0, the velocity of the piston. The integral is taken on the surface of the piston S and r' is the distance between the observation point at point (x, y, z) and the surface element $ds = dx' \, dy'$ at point $(x', y', 0)$,

$$r' = [(x - x')^2 + (y - y')^2 + z^2]^{1/2} \quad \text{(m)} \tag{4.3.21}$$

The integral of Eq. (4.3.20) is obtained analytically for a large r as

$$P(r, \theta) = \frac{j\rho c k a^2 V_0}{2r} \, e^{-jkr} [2J_1(ka \sin \theta)/(ka \sin \theta)] \quad \text{(Pa)} \tag{4.3.22}$$

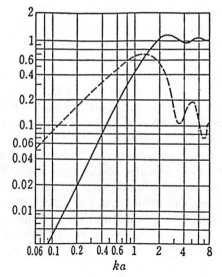

Fig. 4.3.8. Radiation impedance (real: solid, imaginary: broken) of a circular flat piston with radius a in an infinite baffle. From Kido (1988).

where a is the radius of the circular disk, r is the distance from the centre of the piston to the observation point, θ is the gazing angle from the piston axis to the observation point and $J_1(x)$ is the first-order Bessel function of the first kind. Equation (4.3.22) gives the directivity of the piston source, $D(\theta)$, in the far field.

The integral of Eq. (4.3.20) in the near field is more complex and the solution has been given analytically by infinite series of spherical Bessel and Legendre functions. Recently, however, a numerical integration of Eq. (4.3.20) is becoming more common since fast and small computers are easily available. Some examples of the pressure distributions in front of the piston are shown in Fig. 4.3.7 (Kido, 1988). In the low frequency range, the circular piston radiates like a monopole source. But as the frequency increases, the radiation becomes more directional. Note also that the pressure is finite on the surface of the piston.

The self radiation (mechanical) impedance of the circular piston is also known. It is given by

$$Z(ka) = \pi a^2 \rho c \left(1 - \frac{J_1(2ka)}{ka} + \frac{jS(2ka)}{ka} \right) \quad (\text{Ns/m}) \qquad (4.3.23)$$

where $S(x)$ is the Struve function. For small ka, it is approximated by

$$Z(ka) = \pi a^2 \rho c \left(\frac{(ka)^2}{2} + \frac{j8(ka)}{3\pi} \right) \qquad (4.3.24)$$

The radiation impedance normalized to $\pi a^2 \rho c$ is shown in Fig. 4.3.8 (Kido, 1988). The radiation resistance is proportional to f^2 and the reactance is proportional to f in the frequency range $ka \leqslant 1.0$.

4.3.4. Convex and Concave Domes in an Infinite Baffle

A sound radiation problem from a flat source in an infinite baffle is simple even if the source shape and its velocity distribution are complex. The solution is given by the Huygens–Rayleigh integral (Eq. (4.3.20)), which is easily integrated numerically. However, most common loudspeaker diaphragms are not flat. An interesting question is whether there is any difference in the sound radiation characteristics between flat, convex and concave radiators. Differences of radiation impedances and directivity patterns of the three source types are discussed here.

The sound radiation from a convex or a concave source in an infinite baffle is much more complex. Several methods are applicable to these problems. The finite element method is suitable for a problem within a finite space, whereas the boundary element method can be used for both finite and infinite space problems. There are some other special techniques that are suitable for indivdual problems. Sound radiation problems from axisymmetric convex and concave sources in an infinite baffle can be solved by a semianalytic method (Suzuki, 1981a, b). Results obtained by this method are introduced here.

A. *Radiation from a Convex Dome*

For the case of a convex dome in an infinite baffle, the radiation is equivalent to the radiation from two convex domes in an infinite free space with their backs connected together as shown in Fig. 4.3.9. The velocity potential outside the sphere with its centre

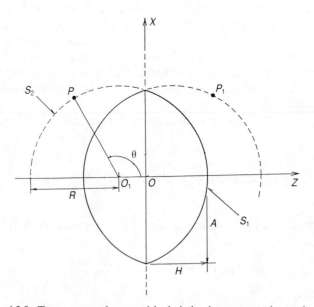

Fig. 4.3.9. Two convex domes with their backs connected together.

at O_1 and with radius R is expressed in terms of spherical wave functions such as

$$\phi(r, \theta) = \sum_{n=0}^{\infty} a_n h_n^{(2)}(kr) P_n(\cos \theta) \qquad (4.3.25)$$

where a_n is an unknown coefficient that must be determined from the boundary conditions, $h_n^{(2)}(kr)$ is the nth-order spherical Hankel function of the second kind, and $P_n(\cos \theta)$ is the nth-order Legendre function. On the surface S_1 of the right-side dome, the particle velocity is prescribed (Neumann-type boundary condition). On the imaginary surface S_2 of the sphere, the condition that the velocity potential at point P is equal to that at point P_1 (image of P) can be used (modified version of Dirichlet-type condition). With these mixed boundary conditions, the unknown coefficients are determined by use of the least-square error method.

B. Radiation from a Concave Dome

In case of a concave dome as shown in Fig. 4.3.10, the velocity potential inside the sphere is expressed in a similar form as Eq. (4.3.25) with the origin at O_1.

$$\phi(r, \theta) = \sum_{n=0}^{\infty} a_n j_n(kr) P_n(\cos \theta) \qquad (4.3.26)$$

where $j_n(kr)$ is the nth-order spherical Bessel function. The same condition (prescribed velocity potential) is given on the radiator surface S_1. For the potential matching, the particle velocity distribution normal to the baffle surface on the opening of the cavity is obtained using Eq. (4.3.26). Then anywhere in the semi-infinite space the velocity

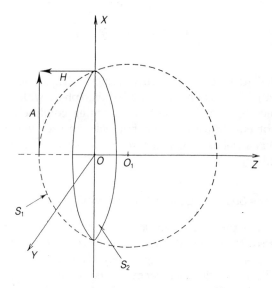

Fig. 4.3.10. Concave dome in an infinite baffle.

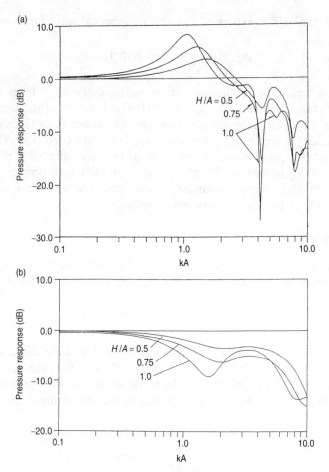

Fig. 4.3.11. Far-field on-axis sound pressure response of convex (a) and concave (b) domes normalized to that of a flat piston.

potential is calculated from the normal velocity distribution on the opening surface using Eq. (4.3.20). The imaginary surface of the sphere with the origin at O_1 seems to be a good choice to give the potential matching condition. However, this causes a convergence problem. Instead, the opening surface itself was used. The choice of this surface is not very strict as long as the imaginary surface S_2 or any other surfaces close to it are avoided.

C. Far-field On-axis Sound Pressure Characteristics

The far-field on-axis pressure responses of the convex and concave domes normalized to the sound pressure of a point source at the origin O with the same volume velocity are shown in Fig. 4.3.11a and b, respectively. The zero-decibel line is the response of the flat piston. There are clear differences in the on-axis pressure characteristics among the flat piston and the convex and concave domes. The concave dome has a peak around $kA = 1.0$ due to the cavity. On the other hand, the on-axis sound pressure response of the convex dome is always lower than that of a flat piston.

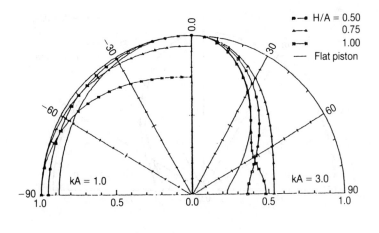

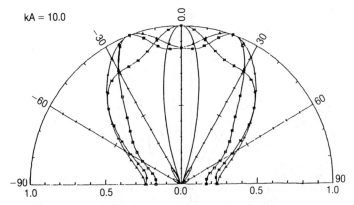

Fig. 4.3.12. Directivity patterns of a convex dome.

D. *Directivity Patterns*

The directivity patterns of the flat piston and the convex dome for $kA = 1, 3$ and 10 are shown in Fig. 4.3.12 and those of the concave dome for $kA = 1, 3, 4.16$ and 10 are shown in Fig. 4.3.13. The normalized frequency $kA = 4.16$ corresponds to the deep trough of the on-axis pressure characteristic shown in Fig. 4.3.11(b). The convex and concave domes have wider directivity patterns than the flat piston. It is interesting that the convex dome with $H/A = 0.75$ and 1.0 has a pressure minimum on the axis ($\theta = 0$).

E. *Intensity Distributions around Radiators*

The reasons why the convex and concave domes have wider directivity patterns than the flat piston are understood by visualizing the sound field around the radiators. Figures 4.3.14 and 4.3.15 show the intensity vectors and pressure distributions of the convex and concave domes. The convex dome cannot keep the energy close to the axis because of its shape. The concave dome concentrates the energy at the focus point and then the energy spreads out again. Intensity vectors at $kA = 4.15$ show a vortex, which may be explained as a similar phenomenon to that indicated in Section 4.2.2.B.

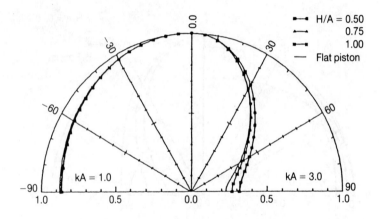

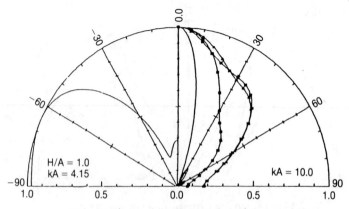

Fig. 4.3.13. Directivity patterns of a concave dome.

4.3.5. Loudspeakers

There are many cases in which more than one loudspeaker is used, such as in sound reproduction and active noise control. In a multiway loudspeaker system, different loudspeaker units for the low-, mid-, and high-frequency ranges are used. For stereo sound reproduction, two or more loudspeakers are used in a room. For active noise control, a secondary sound source is added. If there is more than one loudspeaker, interaction occurs among them. This interaction mechanism can be visualized by use of the intensity method.

Figure 4.3.16 shows the sound intensity and the pressure distribution in front of two closely located loudspeakers. The two loudspeakers are driven 180° out of phase and the drive voltage ratio is 1:0.5 (right vs left). The arrows show the intensity vectors and the contour lines show the pressure levels. Figure 4.3.16 clearly shows the absoprtion of energy by the left loudspeaker. This is an experimental confirmation of the result given in Fig. 4.3.1c.

Since the self and mutual radiation resistances determine the total radiated power, it is very important to measure those values. This is possible by measuring $|v_1|$, $|v_2|$, θ and W_t in Eq. (4.3.9) for three independent drive conditions and solving a set of

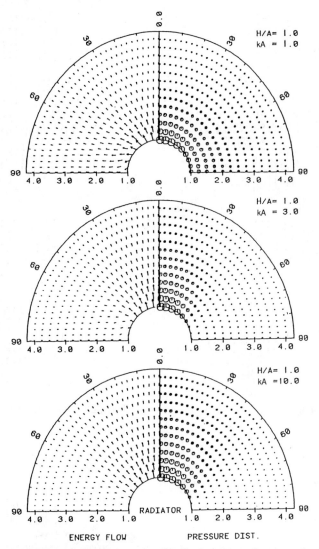

Fig. 4.3.14. Intensity vector and pressure distribution around convex domes.

simultaneous equations with respect to R_{11}, R_{22} and R_{12} (see Section 6.1.2). From measured radiation resistances, it is possible to estimate roughly the largest power reduction by use of Eq. (4.3.18) if it is assumed that source 1 is not affected by the drive condition of source 2.

4.3.6. Rectangular Plates

Since panel vibrations are common sound sources in the field of mechanical vibration and room acoustics, it is important to understand the vibration and sound radiation mechanisms from them. Visualization of the vibration patterns and the sound radiation helps us better understand those phenomena.

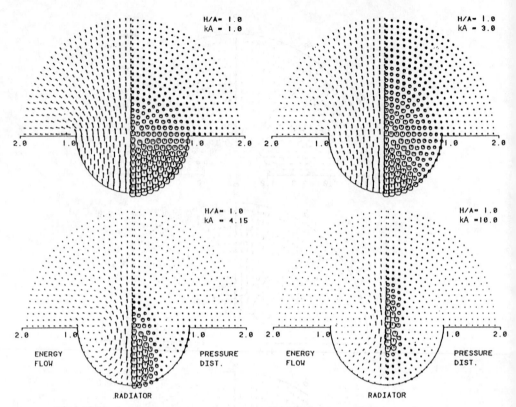

Fig. 4.3.15. Intensity vector and pressure distribution around concave domes.

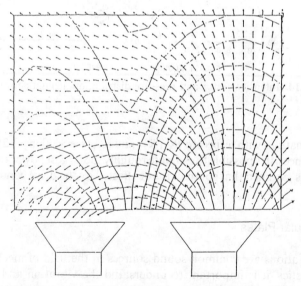

Fig. 4.3.16. Sound intensity vector and pressure contour in front of two closely located loudspeakers at 300 Hz.

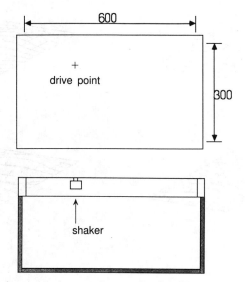

Fig. 4.3.17. A stainless steel plate used to study vibration and radiation characteristics.

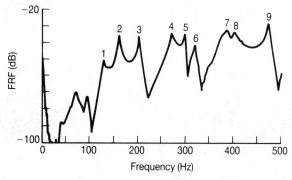

Fig. 4.3.18. A frequency response function between two points on the plate.

A stainless-steel plate shown in Fig. 4.3.17 was used to study the relation between the vibration and radiation characteristics. The modal analysis technique was used to obtain the modal shapes at several resonances. The plate was driven by a small shaker at the point marked + and the vibration was measured at 153 points all over the plate. The intensities normal to the plate surface were also measured on a parallel plane 50 mm above the plate. One of the frequency response functions (acceleration/driving force) is shown in Fig. 4.3.18. It shows nine resonances of the plate in the frequency range up to 500 Hz. Two small resonances below 100 Hz are due to the rigid motion of the plate–frame system. The modal shapes are shown in Fig. 4.3.19. Since the plate shape is rectangular, the modal shapes are very simple except for high-order modes. It seems that the mass of the shaker affects the modal shapes of high orders.

The intensity patterns at resonance frequencies are shown in Fig. 4.3.20. The nodal lines of the intensity patterns and the modal shapes of the low-order modes coincide very well. Therefore, intensity patterns measured close to a radiator surface can be used

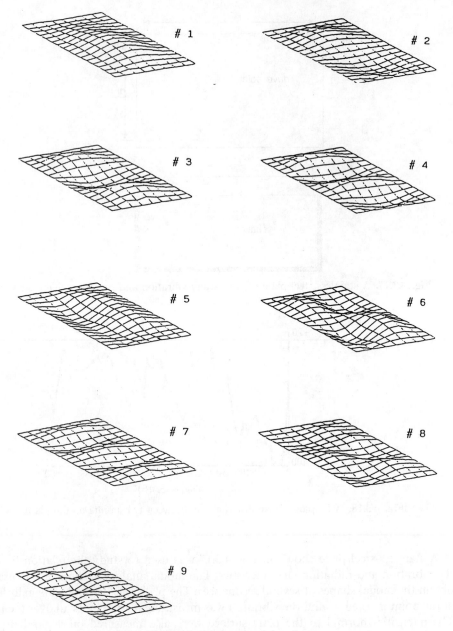

Fig. 4.3.19. Modal shapes of the plate.

to estimate roughly the vibration patterns. When Figs 4.3.19 and 4.3.20 are compared, one should recall that the intensity patterns are time averages of sound radiation due to the forced vibration of the plate, which comprises many modes. On the other hand, the modal shapes are those inherent to individual modes. In the present case, since the damping of the stainless-steel plate is very small, an intensity pattern at one of the resonance frequencies is mostly determined by the mode shape of the specific resonance.

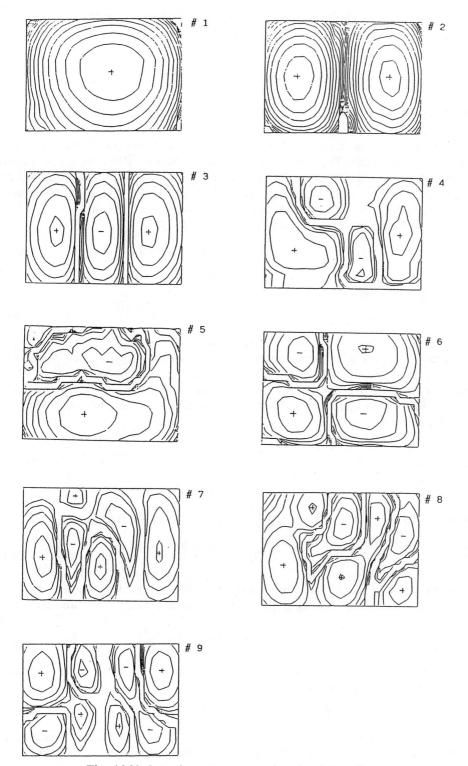

Fig. 4.3.20. Intensity patterns normal to the plate surface.

If the damping is very large or if there is a modal overlap, the nodal lines of the intensity patterns and the mode shapes may be quite different.

4.4. Other Types of Sound Intensities

4.4.1. Envelope Intensity

The property that the time-average sound intensity is independent of time is the reason why it is used widely. There are some occasions, however, when the time-dependent nature of the sound intensity is needed. When a transient sound radiation is investigated, the time-average sound intensity is not convenient. In such cases a slowly varying intensity can be used. The envelope intensity is one of them.

The envelope intensity is defined as (Suzuki et al., 1991)

$$I_e(t) = \tfrac{1}{2}\text{Re}[p_a^*(t)v_a(t)] \quad (\text{W/m}^2) \tag{4.4.1}$$

where $p_a(t)$ and $v_a(t)$ are the analytic functions of $p(t)$ and $v(t)$, respectively. For example, $p_a(t)$ is obtained from $p(t)$ by (see Eq. (2.3.25))

$$p_a(t) = [p(t) - jp_h(t)] \quad (\text{Pa}) \tag{4.4.2}$$

where $p_h(t)$ is a Hilbert transform of $p(t)$. The analytic function $p_a(t)$ has a positive frequency spectrum only. The energy of the analytic function $p_a(t)$ defined by Eq. (4.4.2) is twice the energy of the original function $p(t)$. The division by 2 in Eq. (4.4.1) is necessary in order to maintain the energy equality between the original and the analytic signals. For a sinusoidal wave, defined by Eqs (4.1.3) and (4.1.4), the analytic functions are given, respectively, by

$$p_a(t) = Pe^{j2\pi ft} \tag{4.4.3}$$

and

$$v_a(t) = Ve^{j(2\pi ft + \theta)} \tag{4.4.4}$$

Substituting Eqs (4.4.3) and (4.4.4) into (4.4.1), the envelope intensity is given by

$$I_e(t) = \frac{PV(\cos\theta)}{2} \tag{4.4.5}$$

The envelope intensity of the sinusoidal signal is time-independent. For a time-varying signal, $p_a(t)$ and $v_a(t)$ are represented by use of instantaneous envelopes and phases. For example,

$$p_a(t) = A_p(t)e^{j\theta_p(t)} \tag{4.4.6}$$

where $A_p(t)$ and $\theta_p(t)$ are the instantaneous envelope and the instantaneous phase of $p_a(t)$, respectively. Equation (4.4.1) can be rewritten as

$$I_e(t) = \tfrac{1}{2}A_p(t)A_v(t)\cos(\theta_p(t) - \theta_v(t)) \tag{4.4.7}$$

If the phase difference between $\theta_p(t)$ and $\theta_v(t)$ does not vary much in time, $I_e(t)$ is approximately proportional to the multiple of the instantaneous envelopes of the pressure and the particle velocity.

The analytic signals of the pressure and particle velocity are easily obtained by applying the fast Fourier transforms to the two pressure signals, $p_1(t)$ and $p_2(t)$, of the two-microphone method. The procedure is as follows: (1) Obtain Fourier spectra of the pressure and the particle velocity using Eqs (4.1.18) and (4.1.21). (2) Make the negative frequency components equal to zero. If one wants to analyse a band-limited portion of a signal, one can use filters in this step. (3) Apply the inverse Fourier transform to them. (4) Multiply them and take the real part as defined by Eq. (4.4.1).

The first example of the application of the envelope intensity is to a burst sound generated by a loudspeaker in an anechoic room. The test setup is shown in Fig. 4.4.1. The burst sound is a portion (five cycles) of a 5 kHz sinusoidal wave. The direct sound and the reflected sound from a particle board (900 mm × 900 mm × 20 mm) are received by the two microphones. Figure 4.4.2 shows the result. The top, middle, and the bottom figures are the pressure, instantaneous intensity, and the envelope intensity, respectively. The pressure signal does not have any directional information. The instantaneous intensity shows the direction of the wave packet propagation. However, it varies too rapidly, making observation rather complex. The envelope intensity, on the other hand, is slowly varying as well as having the directional property. Another example is shown in Fig. 4.4.3. The same particle board is hit by an impact hammer as used for the modal analysis at the point shown by an arrow in the figure. Transient sound generated by the impact is measured in x- and y-directions at measurement points a (0 mm, −450 mm) and point b (−450 mm, −450 mm). The upper and lower illustrations in Fig. 4.4.3 show continuously changing directions and magnitudes of the envelope intensity and the instantaneous intensity, respectively. The upper figure clearly shows that the largest sound energy flow is from the direction of the impact. The instantaneous intensity shows the direction of the impact but it varies too fast, making the phenomenon look very complex. As the above examples show,

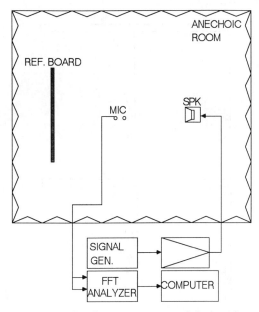

Fig. 4.4.1. Test set-up for the measurement of the envelope intensity.

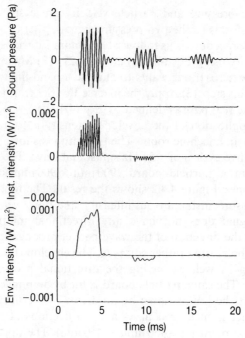

Fig. 4.4.2. Pressure (top), instantaneous intensity (middle), and envelope intensities (bottom) of the direct and reflected sound by a particle board.

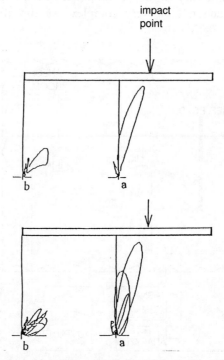

Fig. 4.4.3. Envelope (top) and instantaneous (bottom) intensities of a transient impact sound.

the envelope intensity can be used to visualize the magnitude and direction of a transient sound.

4.4.2. Instantaneous Sound Intensity Spectrum

The envelope intensity gives a time-varying intensity. However, the frequency resolution is limited by the filter bandwidth. If the filter bandwidth is too narrow, the filter response may be longer than the signal response, making the analysis erroneous. One solution of this kind of problem is to use a method that can represent a signal on the time–frequency $(t-f)$ plane. The Wigner distribution is considered as a method that represents signal energy on the $t-f$ plane (see Section 2.4.3); this method can be used as a tool for analysing a time-dependent intensity spectrum.

The cross Wigner distribution between $p(t)$ and $v(t)$ is defined as (see Eq. (2.4.22))

$$W_{pv}(t, f) = \int_{-\infty}^{\infty} p^*(t - \tau/2)v(t + \tau/2)e^{-j2\pi f\tau}\, d\tau \qquad (4.4.8)$$

If $p(t)$ and $v(t)$ are given by Eqs (4.1.18) and (4.1.21), respectively, a time-dependent intensity spectrum can be defined by

$$I(t, f) = -\frac{\mathrm{Im}[W_{p1p2}(t, f)]}{2\pi fpd} \qquad (4.4.9)$$

which will be called the ISIS (instantaneous sound intensity spectrum). ISIS makes it possible to analyse the intensity spectrum of a transient sound on the $t-f$ plane (Kawaura et al., 1987).

Figure 4.4.4 shows geometries (in millimetres) of a midrange and a tweeter loud-speaker in an enclosure and a measurement point. A chirp signal is given to the loudspeaker system and the sound at the measurement point is measured in the x- and y-directions. Figure 4.4.5 shows the ISIS vector on the $t-f$ plane. The waveform along the time axis shows one of the two pressure signals measured by the two microphones.

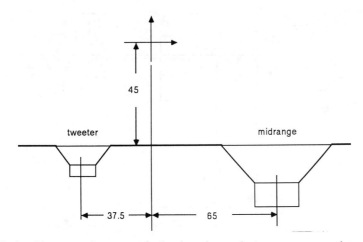

Fig. 4.4.4. A midrange and tweeter of a loudspeaker and a measurement point.

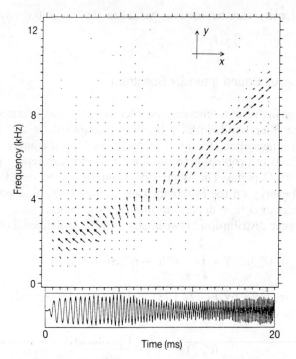

Fig. 4.4.5. Instantaneous sound intensity spectrum measured in front of a loudspeaker.

The figure clearly shows that the sound comes from the midrange to the measurement point at the beginning and then from the tweeter in the high-frequency range. This transition occurs at around 8 ms and 3 kHz.

Another experimental set-up is shown in Fig. 4.4.6. Three particle boards simulate three walls (ceiling, side wall and floor). A loudspeaker radiates a sound (4 cycles of a 10-kHz sinusoidal wave). The ISIS vectors are shown in Fig. 4.4.7. The first and largest

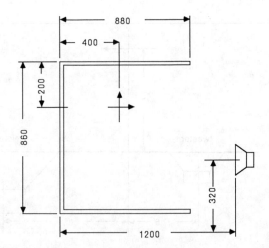

Fig. 4.4.6. An experimental set-up simulating three walls.

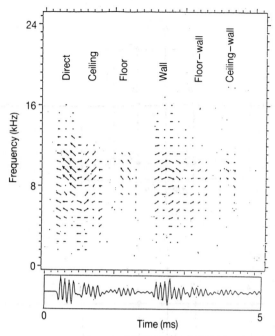

Fig. 4.4.7. Instantaneous sound intensity spectrum of direct and reflected sound by the three walls.

wave packet is the direct sound from the loudspeaker. This is confirmed by the vector display. The energy flows from the lower right to the upper left. The gravitational centre of the spectrum is at around 10 kHz but is widely distributed. The second wave packet is the first reflected sound from the ceiling. The following wave packets are also identifiable by the wall from which they are reflected.

One thing we must consider in interpreting the Wigner distribution is the existence of interferences. For example, we can observe vectors flowing from left to right between the direct and the first reflected sound. This is due to the interference between those two wave packets. If there are two signals separated in time, interference occurs between them. Also, if there are two spectra separated in frequency then there is interference between them. Another difficulty in the interpretation of the Wigner distribution is the existence of negative values. If the Wigner distribution is considered as the energy density on the t–f plane, it must be positive definite in the conventional sense of energy. However, negative values appear very often on the t–f plane even if they satisfy the conditions of Eqs (2.4.24) and (2.4.25). Other types of distributions are being investigated to avoid negative densities on the t–f plane (Choi and Williams, 1989).

4.4.3. Vibration Intensity

Since the sound is mostly radiated by vibration of a structure, it is very important to know how the vibration is transmitted in the structure. The vibration intensity (or structural intensity) is defined as the power flow through a unit width of a structure. It

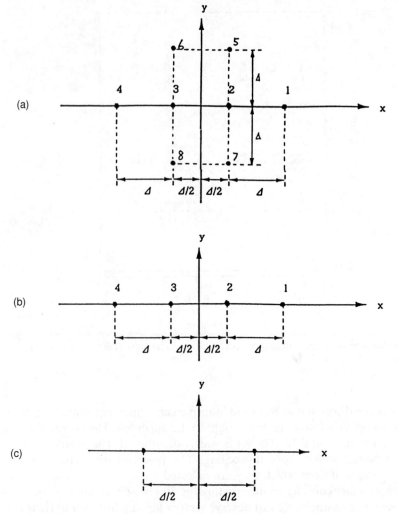

Fig. 4.4.8. Eight-point, four-point and two-point measurement methods of vibration intensity.

has similarities with as well as differences from the sound intensity. It can be used to visualize the energy flow and to find sources, paths and sinks of the vibrational energy. The vibration intensity can be measured by a method very similar to that for the sound intensity. One significant difference is that the structure is more complex than the air as a medium of energy propagation. The properties of the air can be represented simply by the density and the sound velocity, which are generally constant and direction-independent. On the other hand, a structure has many physical constants as well as physical dimensions.

The derivation of the vibration intensity has been given by Noiseux (1970) and Pavic (1976, 1981), and it will be briefly described here. For simplicity a bending wave of a plate of constant thickness is considered. The instantaneous vibration intensity in the

x-direction of a plate on the x–y plane is given by

$$I_x(x, y, t) = I_x^{sf} + I_x^{bm} + I_x^{tm} \quad \text{(W/m)} \tag{4.4.10}$$

where I_x^{sf}, I_x^{bm} and I_x^{tm} are vibration intensities due to the shear force, bending moment, and torsional moment, respectively. These are given by

$$I_x^{sf} = B \frac{\partial}{\partial x}\left(\frac{\partial^2 \zeta}{\partial x^2} + \frac{\partial^2 \zeta}{\partial y^2}\right)\frac{\partial \zeta}{\partial t} \tag{4.4.11}$$

$$I_x^{bm} = -B\left(\frac{\partial^2 \zeta}{\partial x^2} + \frac{\partial^2 \zeta}{\partial y^2}\right)\frac{\partial^2 \zeta}{\partial x\,\partial t} \tag{4.4.12}$$

$$I_x^{tm} = -B(1-v)\frac{\partial^2 \zeta}{\partial x\,\partial y}\frac{\partial^2 \zeta}{\partial y\,\partial t} \tag{4.4.13}$$

where B is the bending stiffness, ζ is the displacement in the z-direction, and v is Poisson's ratio. In order to approximate spatial differentiations by linear differences in the 2-dimensional space, displacements at 8 points as shown in Fig. 4.4.8a must be known. If the vibration is stationary, the time-average spectrum of $I_x(x, y, t)$ is given as

$$I_x(x, y, f) = \frac{2\pi f B}{4\Delta^3} \{6(4 + v)\mathrm{Im}[G_{36}] - 4(\mathrm{Im}[G_{16}] - \mathrm{Im}[G_{38}])$$

$$+ (1 - v)(-2\mathrm{Im}[G_{23}] + \mathrm{Im}[G_{25}]) - \mathrm{Im}[G_{27}] + 2\mathrm{Im}[G_{34}]$$

$$- \mathrm{Im}[G_{45}] + \mathrm{Im}[G_{47}] + 2\mathrm{Im}[G_{56}] - 2\mathrm{Im}[G_{67}]$$

$$+ 2(1 + v)(-\mathrm{Im}[G_{26}] - \mathrm{Im}[G_{35}] - \mathrm{Im}[G_{37}] - \mathrm{Im}[G_{46}])\} \tag{4.4.14}$$

where G_{ij} is the one-sided cross spectrum between the ith and jth point accelerations and Δ is the separation between measurement points.

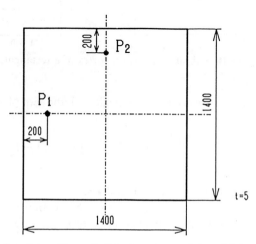

Fig. 4.4.9. Rectangular plate with negligible damping and two driving point locations.

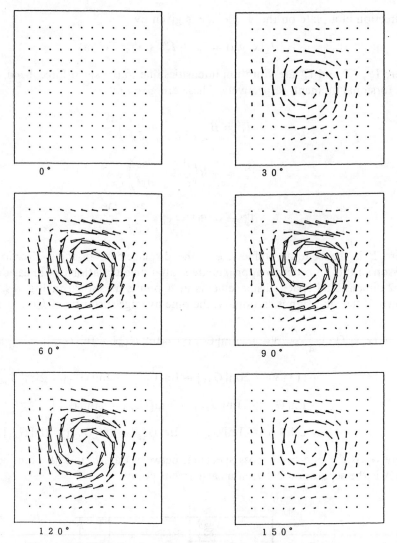

Fig. 4.4.10. Vibration intensity vortex of a rectangular plate.

If the vibration is 1-dimensional, I_x^{tm} is neglected and Eq. (4.4.14) is simplified as

$$I_x(x, y, f) = \frac{2\pi f B}{\Delta^3} \{4\text{Im}[G_{23}] - \text{Im}[G_{24}] - \text{Im}[G_{13}]\} \qquad (4.4.15)$$

The measurement points are shown in Fig. 4.4.8b.

If the vibration is 1-dimensional and if the observation point is far from either a driving point or a discontinuity, a further simplification is possible such as

$$I_x(x, y, f) = \frac{(2\pi f)^2 \sqrt{B\rho}}{\Delta} \text{Im}[G_{12}] \qquad (4.4.16)$$

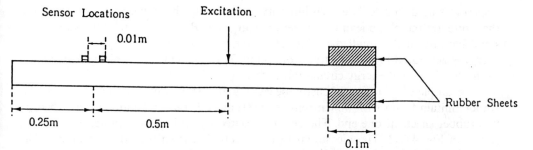

Fig. 4.4.11. Aluminium beam used for the experiment.

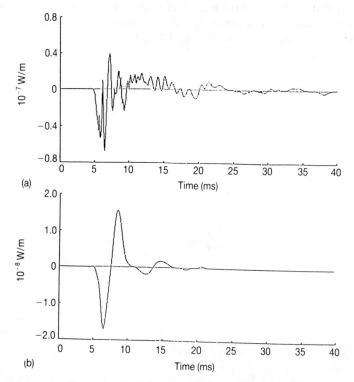

Fig. 4.4.12. (a) Overall and (b) band-passed envelope intensities.

where ρ is the density per unit area. The measurement points are shown in Fig. 4.4.8c. As Eq. (4.4.16) indicates, there is a high degree of analogy between the sound and the vibration intensities.

Figure 4.4.9 shows a model plate used for the numerical investigation of the vibration intensity. The plate size is 1400 mm × 1400 mm × 5 mm. The Young's modulus and the density of the plate are 2.02×10^{11} N/m² and 7860 kg/m³, respectively. The plate has a small structural damping (3%). The driving points are P_1 and P_2 in the figure. The frequency is 92 Hz, which is equal to the $(1, 0)$ and $(0, 1)$ mode frequency. Figure 4.4.10 shows numerically obtained energy flows when the plate is driven with various phase differences at the two points. The eight-point method (Fig.

4.4.8a) was used to calculate each intensity component. Phase differences (θ) shown in the figure are the phase lead of driving force at point P_1. If $\theta = 0$, the intensity is very small because of the small damping. However, as θ increases, a large circulation of energy is generated in the direction from point P_1 to P_2. This is a similar phenomenon to that in the sound energy circulation described in Section 4.2.

An example of vibration intensity measurement is shown by Figs 4.4.11 and 4.4.12. An aluminium beam with dimensions of 1000 mm × 30 mm × 2 mm is sandwiched by two rubber sheets at one end. The beam is excited by an impact hammer at the point shown in Fig. 4.4.11. The measurement point is 0.25 m from the other (free) end. The separation of the two sensors is 10 mm. The measured vibration (envelope) intensity below 5 kHz is shown by Fig. 4.4.12a. At the beginning, the energy flows from the right to the left and, in a very short period of time, it is reflected at the free end and comes back to the measurement point. When an octave filter centred at 1 kHz is used, this phenomenon becomes very clear (Fig. 4.4.12(b)).

5

Subjective and Physiological Responses to Sound Fields

Instead of dealing with the peripheral behaviour of the auditory system and its related subjective attributes, this chapter uses the so-called 'top-down' approach. It first describes the theory of the subjective attributes in relation to four orthogonal physical factors, including the temporal and spatial factors of a sound field. Then it analyses auditory evoked potentials that may relate to a primitive subjective response, the subjective preference. According to such fundamental phenomena, a workable model of the human auditory–brain system is proposed, and this model well describes any subjective attribute—including loudness, subjective diffuseness, and even speech articulation.

As an application of this model, the four orthogonal factors are used to determine the preferred condition for each listener. After construction of the concert hall, a seat selection system for each listener, testing his/her own subjective preference, may be introduced. Basic subjective attributes such as loudness, coloration, the threshold of perception of a reflection, and echo disturbance can be mainly described in relation to the autocorrelation function (ACF) of the source signals. Subjective diffuseness as well as speech clarity and intelligibility are described by the IACC, in addition to the temporal factors.

5.1. Theory of Subjective Preference and the Limitation of Orthogonal Acoustic Factors

Research on the perplexing problems that arise in the design of concert halls was conducted in response to the question of how best to precisely identify and describe the design qualities. Different source programs (reverberation-free speech and music) and a computer are used to simulate the sound fields of signals arriving at both ears, and the law of comparative judgements enables us to attain the linear scale value or the distance between sound fields for certain subjective responses. Four independent factors contributing to good acoustics are thus described. The design theory derived here makes it possible to calculate the acoustical quality at any seat in a proposed concert hall, enabling musicians to choose music most suited to a performance in a particular concert hall.

5.1.1. Physical Description of a Sound Field and the Limitation of Orthogonal Factors

To simplify the analysis, first consider a single sound source in a room. Let $h_L(r \mid r_o; t)$ and $h_R(r \mid r_o; t)$ be the impulse responses between the source point r_o and the left and right ear-canal entrances of a listener whose head is centered at $r(x, y, z)$. Then the pressures at both ears, which must include all acoustic information to be analysed, are expressed by

$$f_{L,R}(t) = p(t) \otimes h_{L,R}(r \mid r_o; t) \tag{5.1.1}$$

where $p(t)$ is a source signal and $\otimes$ denotes convolution. The impulse responses may be decomposed into a set of impulse responses $w_n(t)$ describing the reflection property of boundaries and the impulse responses $h_{nL}(t)$ or $h_{nR}(t)$ from the free field to the ear-canal entrance, n denoting a single sound with a horizontal angle ξ to the listener and with an elevation angle η ($\xi = 0$ and $\eta = 0$ signify the frontal direction). The impulse responses may now be written:

$$h_{L,R}(r \mid r_o; t) = \sum_{n=0}^{N} A_n w_n(t - \Delta t_n) \otimes h_{nL,R}(t) \tag{5.1.2}$$

where A_n is the pressure amplitude and Δt_n is the delay time of the reflections relative to the direct sound. The amplitude A_n is determined by the '(1/r)-law', A_0 being unity. Every $n (\geqslant 1)$ corresponds to a single reflection, and $n = 0$ refers to the direct sound. Equation (5.1.1) thus becomes

$$f_{L,R}(t) = \sum_{n=0}^{N} p(t) \otimes A_n w_n(t - \Delta t_n) \otimes h_{nL,R}(t) \tag{5.1.3}$$

If the sound source radiates nonuniformly, the radiation pattern of the sound source is taken into consideration for each sound direction—that is, $p(t)$ in this equation is replaced by $p_n(t)$. When there are many sound sources distributed on a stage, for usual levels of sound pressure the pressures at the two ears may be expressed as a linear sum of the $f_{L,R}(t)$ values given by Eq. (5.1.3).

All independent objective parameters of the acoustic information, which are included in the sound pressures given by Eq. (5.1.3), may be reduced to the following limited number of orthogonal or independent factors (Ando, 1983).

1. The first parameter is the autocorrelation function (ACF) of the source signal $p(t)$ defined in Section 2.2.3:

$$\Phi_p(\tau) = \frac{1}{2T} \int_{-T}^{+T} p'(t) p'(t + \tau) \, dt$$

Here $p'(t) = p(t) \otimes s(t)$, $s(t)$ corresponds to the ear sensitivity, which might be characterized by the external ear and the middle ear. This function indicates a repetitive feature or a reverberation of the sound source itself. The above equation may be divided by the intensity of the source signal $\Phi_p(0)$ to give the normalized ACF, which is defined by

$$\phi_p(\tau) = \frac{\Phi_p(\tau)}{\Phi_p(0)} \tag{5.1.4}$$

(a)

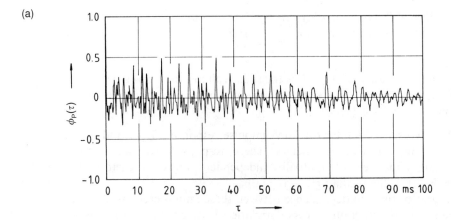

(b)

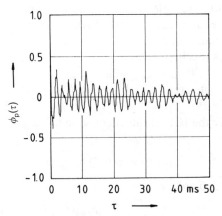

Fig. 5.1.1. Measured ACFs of music. The duration of the music signal is 35 s, and the effective duration of the ACF is defined by the delay τ_e at which the envelope of the normalized ACF becomes 0.1. (a) Music motif A, Royal Pavane by Gibbons, $\tau_e = 127$ ms. (b) Music motif B, Sinfonietta, Opus 48, III movement by Malcolm Arnold, $\tau_e = 43$ ms. From Ando (1983).

Examples of measured ACF ($2T = 35$ s) are shown in Fig. 5.1.1 (temporal–monaural criterion).

2. The second objective parameter is the set of impulse responses of the reflecting boundaries, $A_n w_n(t - \Delta t_n)$, which represents the initial time delay Δt_1 between the direct sound and the first reflection as well as the subsequent reverberation time and any spectral changes due to the reflections. Note that there are certain relationships between Δt_n and A_n, and Δt_1 and $\Delta t_n (n \geqslant 1)$ because the density of reflections statistically increases with increasing the delay time squared. At the design stage, the subsequent reverberation may be approximated by the conventional Sabine formula for reverberation time:

$$T_{\text{sub}} \approx \frac{KV}{\bar{\alpha}S} \quad \text{(s)}$$

where K is a constant (≈ 0.161), V is the room volume, (m^3), $\bar{\alpha}$ is the average absorption coefficient, and S (m^2) is the total area of the room surface (temporal–monaural criterion).

3. The two sets of the head-related impulse responses, $h_{nL}(t)$ and $h_{nR}(t)$, constitute the remaining objective parameters. These responses play an important role in localization, and are not mutually independent objective factors. For example, $h_{nL}(t) \approx h_{nR}(t)$ in the median plane ($\xi = 0$).

To represent the interdependence between these impulse reponses, we can introduce a single factor: the interaural cross correlation between the continuous sound signals $f_L(t)$ and $f_R(t)$. This factor becomes significant in determining the degree of subjective diffuseness (Ando and Kurihara, 1986). Subjective diffuseness, or no special directional impression, is the perception for sound fields that have a low magnitude of interaural cross correlation. On the other hand, a well-defined direction is indicated when the interaural cross correlation function (ICF) has a strong peak for $|\tau| < 1\,\text{ms}$. If the peak is observed at $\tau = 0$, for example, then a frontal source direction can usually be perceived. The ICF depends mainly on the direction from which reflections arrive at the listener and on their amplitudes.

The ICF is defined as (Section 2.2.3)

$$\Phi_{LR}(\tau) = \frac{1}{2T} \int_{-T}^{+T} f_L(t)\, f_R(t + \tau)\, dt \tag{5.1.5}$$

First let us consider only the ICF of the direct sound. The pressures at the two ears are then expressed by

$$f_{L,R}(t) = p(t) \otimes h_{0L,R}(t)$$

The normalized ICF is defined by

$$\phi_{LR}(\tau) = \frac{\Phi_{LR}(\tau)}{[\Phi_{LL}(0)\Phi_{RR}(0)]^{1/2}} \tag{5.1.6}$$

where $\Phi_{LL}(0)$ and $\Phi_{RR}(0)$ are the ACFs at $\tau = 0$ for each ear. If the listener is facing the source, it approaches unity because $f_L(t) \approx f_R(t)$.

If discrete reflections $n = 1, 2, \ldots, N$ are added to the direct sound after the ACF of the direct sound becomes weak enough, the normalized interaural cross correlation is expressed by

$$\phi_{LR}^{(N)}(\tau) = \frac{\sum_{n=0}^{N} A_n^2 \Phi_{LR}^{(n)}(\tau)}{[\sum_{n=0}^{N} A_n^2 \Phi_{LL}^{(n)}(0) \sum_{n=0}^{N} A_n^2 \Phi_{RR}^{(n)}(0)]^{1/2}} \quad \text{when } w_n(t) = \delta(t) \tag{5.1.7}$$

Here $\Phi_{LR}^{(n)}(\tau)$ is the ICF of the nth reflection, $\Phi_{LL}^{(n)}(0)$ and $\Phi_{RR}^{(n)}(0)$ are the ACFs at $\tau = 0$ of the nth reflection at the ears, and $\delta(t)$ is the Dirac delta function.

Since these correlations between the two ears have not been obtained theoretically, the ICF ($2T = 35\,\text{s}$) was measured for each sound arriving at a dummy head. This dummy head was constructed according to an acoustical measurement of the threshold level, so that the output signals of the microphones corresponded to the ear sensitivity.

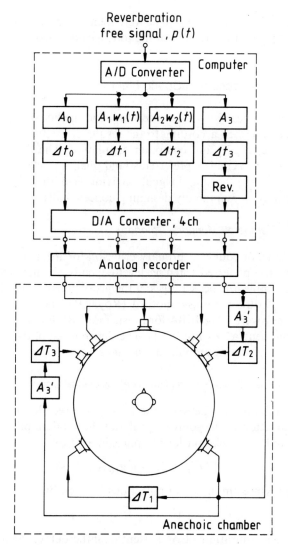

Fig. 5.1.2. A system for simulating sound fields. From Ando (1985), with permission from Springer-Verlag.

These values, which are used to calculate the ICF from Eq. (5.1.7), are listed by Ando (1985).

The magnitude of the ICF is defined by

$$\text{IACC} = |\phi_{LR}(\tau)|_{\max} \qquad \text{for } |\tau| < 1\,\text{ms} \qquad \text{(spatial–binaural criterion)} \qquad (5.1.8)$$

In accordance with Eq. (5.1.3), the sound fields in a concert hall may be simulated by a system as shown in Fig. 5.1.2. A reverberation-free signal is fed through an analogue-to-digital converter into a computer whose program provides the amplitude and the delay of two early reflections ($n = 1, 2$) and the subsequent reverberation ($n \geqslant 3$) relative to the direct sound. The notation 'Rev.' in the figure stands for reverberator. To produce reverberation with 'subjective diffuseness', discrete time delays of ΔT_1, ΔT_2, ΔT_3 were inserted and signals were then fed to the loudspeakers. The IACC was adjusted by varying the location of the loudspeakers around the listener's head.

5.1.2. Subjective Preference of a Sound Field

The optimum design objectives for a concert hall and the linear scale value of subjective preference may be described according to the results of listening tests (paired-comparison tests (Thurstone, 1927)) and computer simulation. The design objectives can be described in terms of subjectively preferred sound qualities, related to the temporal and spatial factors describing the sound signals arriving at both ears. They clearly lead to comprehensive criteria for optimally designing concert halls.

A. *Listening Level (Temporal–Monaural Criterion)*

The listening level is, of course, a primary criterion for listening to the sound field in concert halls, and the preferred listening level depends on the particular music being performed. For example, the peak ranges of the gross preferred levels obtained from 16 subjects were 77–79 dBA for music motif A (Royal Pavane by Gibbons) with a slow tempo, whereas they were 79–80 dBA for music motif B (Sinfonietta by Arnold) with a fast tempo (see different ACFs Fig. 5.1.1). The dependence of the scale values on each objective parameter is shown graphically in Fig. 5.1.3.

B. *Early Reflection After Direct Sound (Temporal–Monaural Criterion)*

An approximate relationship between the most preferred delay time $[\Delta t_1]_p$ and the autocorrelation function of the source signal and the total amplitude of the reflections has been discovered. This relationship is approximated by

$$[\Delta t_1]_p \approx (1 - \log_{10} A)\tau_e \quad \text{(s)} \qquad (5.1.9)$$

where the total pressure amplitude of reflections is given by

$$A = (A_1^2 + A_2^2 + A_3^2 + \cdots)^{1/2} \qquad (5.1.10)$$

and τ_e is the effective duration of ACF (tenth centile delay) defined by the delay at which the envelope of the normalized ACF becomes 0.1. For a sound field with a single reflection ($A = A_1 = 1$), the preferred delays were about 128 ms for the slow music (A, Gibbons) and about 32 ms for the fast music (B, Arnold). It is remarkable that these preferred delays are close to the effective durations of the ACF, for music motifs A and B, 127 ms and 35 ms, respectively (see Fig. 5.1.1).

C. *Subsequent Reverberation Time After Early Reflections (Temporal–Monaural Criterion)*

For the flat frequency characteristics of reverberation (one of the preferred conditions), the subsequent reverberation time preferred is simply described by

$$[T_{\text{sub}}]_p \approx 23\tau_e \quad \text{(s)} \qquad (5.1.11)$$

The most preferred reverberation times estimated for each sound source are shown in Fig. 5.1.4. A lecture and conference room must be designed for speech, whereas an opera house and similar theatres must be designed for vocal music. For orchestral

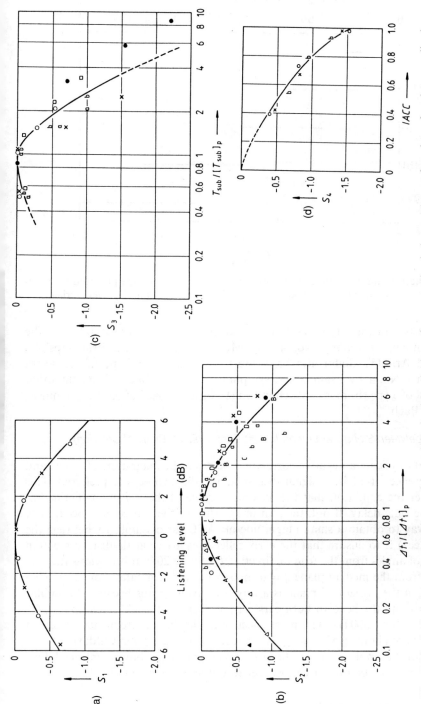

Fig. 5.1.3. Scale values of subjective preference (obtained by the law of comparative judgement) as a function of each normalized factor. Different symbols indicate values obtained in different series of paired-comparison tests with different music motifs. For each factor the scale values of the most preferred conditions are adjusted to zero. (a) Scale values as a function of the normalized listening level. (b) Scale values as a function of the normalized initial time delay between the direct sound and the first reflection. Values of $[\Delta t_1]_p$ may be calculated using Eq. (5.1.9). (c) Scale values as a function of the normalized subsequent reverberation time ($A = 4.1$). Values of $[T_{sub}]_p$ may be calculated by use of Eq. (5.1.11). (d) Scale values as a function of the IACC. The maximum value of ICF must be maintained at $\xi = 0$ to ensure frontal localization of the sound source. The value of IACC may be calculated by use of Eq. (5.1.7) with measured correlation functions for each sound direction. From Ando (1985), with permission from Springer-Verlag.

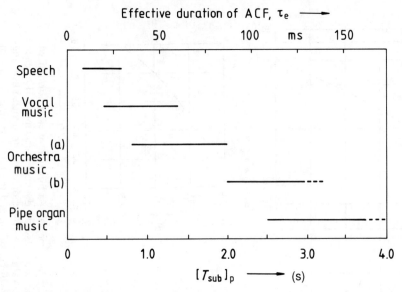

Fig. 5.1.4. The effective durations of ACF for several sound sources, and the preferred reverberation times. Both are estimated. From Ando (1985), with permission from Springer-Verlag.

music, there are two or three types of design, according to the effective duration of the ACF. For example, Symphony No. 41 by Mozart, 'Le Sacre du Printemps' by Stravinsky, and Arnold's Sinfonietta have short ACFs and are Type (a) orchestra music. Symphony No. 4 by Brahms and symphony No. 7 by Buckner, on the other hand are typical of Type (b) music. Much longer ACFs are typical of pipe organ music, such as that of Bach.

D. *Dissimilarity between Signals at Left and Right Ears (Spatial–Binaural Criterion)*

All available data indicate a negative correlation between the magnitude of IACC and subjective preference—that is, a signal dissimilar for the two ears is preferred. This holds only under the condition that the maximum value of the ICF is maintained at the origin of the time delay. If not, then an image shift of the source may occur. The most effective way to obtain a small magnitude of IACC (or dissimilarity between the signals at each ear) is to ensure that the early reflections arrive at the listeners within a certain range of angles from the frontal direction ($\pm 55° \pm 20°$). It is obvious that the sound arriving from the median plane $\xi = 0°$ makes the IACC greater. Sound arriving from $\xi = \pm 90°$ in the horizontal plane is always disadvantageous because the similar 'detour' paths around the head to both ears cannot decrease the IACC effectively, for frequencies higher than 500 Hz. The most effective angles for the frequency ranges of 1 kHz and 2 kHz are about $\pm 55°$ and $\pm 36°$, respectively (Fig. 5.6.9). A fifth significant orthogonal factor has not yet been discovered, so only the above-mentioned four orthogonal factors must be optimized in order to obtain a good sound field.

5.1.3. Theory of Subjective Attributes

Let us now put these results into a theory. Since the number of orthogonal acoustic factors included in the sound signals at both ears is limited as mentioned in Section

5.1.1, the scale value of any 1-dimensional subjective responses may be expressed by

$$S = g(x_1, x_2, \ldots, x_I)$$

This study describes the linear scale value of preference obtained by the law of comparative judgement. Because the four objective parameters have been verified to be affected independently of the scale value and their units are almost constant in the preferred ranges tested, we can add the scale value,

$$S = g(x_1) + g(x_2) + g(x_3) + g(x_4)$$
$$= S_1 + S_2 + S_3 + S_4 \qquad (5.1.12)$$

where S_i is the scale value obtained for each objective parameter. This equation indicates four-dimensional continuity. Note that a fifth dimensionally significant factor is not discovered in usual acoustic spaces. From the nature of the scale value, we may conveniently take a zero value at the most preferred conditions.

Different test series of paired comparison using different music programs yielded the following formula for the scale value of subjective preference (Ando, 1983):

$$S_i \approx -\alpha_i |x_i|^{3/2} \qquad (5.1.13)$$

Here x_1 is given by the sound pressure level difference measured by the A-weighted network, so that

$$x_1 = 20 \log(P/[P]_p) \quad \text{(dB)} \qquad (5.1.14)$$

where P is the sound pressure at a seat and $[P]_p$ is the most preferred sound pressure that may be assumed at a particular seat position in the room under investigation. For the other three parameters,

$$x_2 = \log(\Delta t_1/[\Delta t_1]_p) \qquad (5.1.15)$$

$$x_3 = \log(T_{sub}/[T_{sub}]_p) \qquad (5.1.16)$$

$$x_4 = \text{IACC} \qquad (5.1.17)$$

Thus, the scale values of preference have been approximately formulated in terms of the 3/2 power of the normalized temporal objective parameters expressed in terms of the logarithm. The spatial binaural parameter x_4 is expressed by a term in which it is raised to the 3/2 power of its real value, indicating that its contribution is greater than those of the temporal parameters.

5.1.4. Calculation of Subjective Preference at Each Seat

As a typical example, we shall discuss the quality of the sound field at each seat in a concert hall with a shape similar to that of the Symphony Hall in Boston. It is supposed that a single sound source is located at the centre of the hall, 1.2 m above the stage floor. Receiving points 1.1 m above the floor level correspond to the positions of the ears. The image method is used to take account of reflections and their amplitudes, delay times, and directions of arrival at the listeners. Contour lines of the total scale

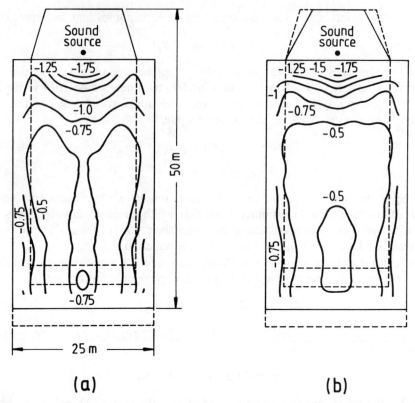

(a)   **(b)**

Fig. 5.1.5. Contour lines of total subjective preference with the four physical factors. (a) The Boston Symphony Hall with the original shape. (b) The Symphony Hall with the side reflector on the stage optimized. From Ando (1985), with permission from Springer-Verlag.

value of preference for music motif B are shown in Fig. 5.1.5a and b. This figure demonstrates that the reflections from the side reflectors on the stage may decrease the IACC values for listeners, thus increasing the preference value at each seat. The reverberation time here is assumed to be 1.8 s throughout the hall, and the most preferred listening level is obtained on the centre line 20 m from the source.

5.2. Slow Vertex Responses (SVR) Relating to Subjective Preference

The previous section discussed four significant and independent physical factors of the sound field in a concert hall. The present section describes efforts to interpret the important qualities of sound in terms of the processes occurring in the auditory pathways and in the brain. If enough were known about how the central nervous system modifies the nerve impulses from the cochlea, concert halls could be designed according to guidelines derived from this knowledge. This subject has been approached by studying the potentials that auditory stimuli evoke in the left and right cerebral hemispheres.

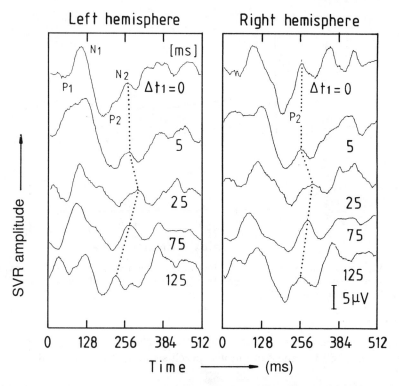

Fig. 5.2.1. Averaged slow vertex responses for a single subject. Dotted lines are the loci against the delay time of the reflection. The upward direction indicates negativity. From Ando et al. (1987a)

5.2.1. Recording Slow Vertex Responses

Each subject was asked to abstain from smoking and from drinking any kind of alcoholic beverage for about 12 hours before testing. To compare slow vertex responses (SVR) with the subjective preference obtained by paired-comparison tests, a reference stimulus was first presented and then an adjustable test stimulus was presented. Such a pair of stimuli was presented alternately 50 times while the SVR was recorded. The electrical responses were obtained from the left and right temporal areas (T_3 and T_4) according to the International 10–20 System (Jasper, 1958). Reference electrodes on the right and left earlobes were connected together.

Examples of the SVR amplitude obtained by averaging the 50 responses for a single subject are shown in Fig. 5.2.1 as a parameter of the delay time of a single reflection. The amplitude of the reflection was the same as that of the direct sound, and the source signal was a 0.9-s fragment of continuous speech 'ZOKI-BAYASHI' (Japanese meaning a grove or a copse). The reference sound field was only the direct sound without any time delay, and the total sound pressure levels in this experiment were kept constant. Two loudspeakers producing the direct sound and the single reflection were located together in front of the subject, so that the magnitude of interaural cross correlation (IACC) could be kept constant at nearly unity for all sound fields tested here. From this figure, we can find that the maximum latency at the most preferred delay time of

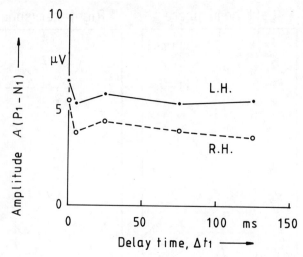

Fig. 5.2.2. Averaged amplitudes of the early slow vertex responses to continuous speech plotted as a function of the delay time of the reflection (8 subjects). LH, left hemisphere; RH, right hemisphere. From Ando et al. (1987a).

reflection is 25 ms, which corresponds well with the effective duration of the autocorrelation function (ACF) of the continuous speech signal (Ando et al., 1987a).

5.2.2. Hemispheric Dominance Depending on Spatial and Temporal Factors

Figure 5.2.2 shows amplitudes of the early SVR defined by P_1-peak to N_1-peak, $A(P_1 - N_1)$, plotted as a function of the delay time of reflection. The amplitude on the left is obviously greater than that on the right (significance level: $p < 0.01$). This may indicate left-hemisphere dominance or specialization of the human brain for such a

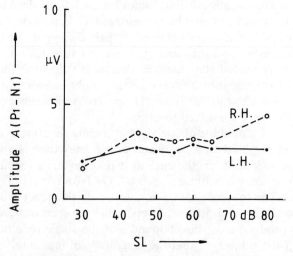

Fig. 5.2.3. Averaged amplitudes of the early slow vertex responses as a function of stimulus intensity (5 subjects). LH, left hemisphere; RH, right hemisphere. From Nagamatsu et al. (1989).

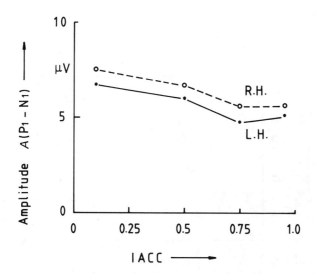

Fig. 5.2.4. Averaged amplitudes of the early slow vertex responses as a function of the IACC (8 subjects). LH, left hemisphere; RH, right hemisphere. From Ando et al. (1987b).

change of the delay time of reflection with speech (see also Table 5.2.1). When the stimulus level was changed, the amplitude of SVR in the right hemisphere was greater than that in the left ($P < 0.01$, except for that at 30 dB sensation level (SL)), even if the speech signal was used as shown in Fig. 5.2.3 (Nagamatsu et al. 1989). When the IACC in the paired stimuli was changed by using 1/3 octave band noise with centre frequency of 500 Hz, the amplitude of the response in the right hemisphere was much greater than that in the left hemisphere (see Fig. 5.2.4, $P < 0.05$) (Ando et al., 1987b).

Summarizing all these data, Table 5.2.1 shows that the hemispheric dominances differed for different sound signals and acoustic factors of the sound field. It is remarkable that hemispheric dominance appeared only in the amplitude of the SVR. For example, even though for the continuous speech signal the right hemisphere was dominant under the condition of varying the SL, the left hemisphere was dominant under the condition of varying the delay time of reflection (which is a temporal factor of the sound field). When the IACC was changed, the right hemisphere was highly activated because of the spatial factor and noise stimulus. It is well known that the left hemisphere is mainly associated with speech and time sequential indentifications, and that the right hemisphere is concerned with nonverbal and spatial identifications

Table 5.2.1. Hemispheric dominances due to the amplitude of the early slow vertex response, $A(P_1 - N_1)$

Source signal	Parameter adjusted	$A(P_1 - N_1)$	p
Speech (0.9 s)	SL	R > L	< 0.01
Speech (0.9 s)	Δt_1	L > R	< 0.01
Speech vowel [a] (see Ando, 1985)	IACC	R > L	< 0.025
1/3-Octave band noise (500 Hz)	IACC	R > L	< 0.05

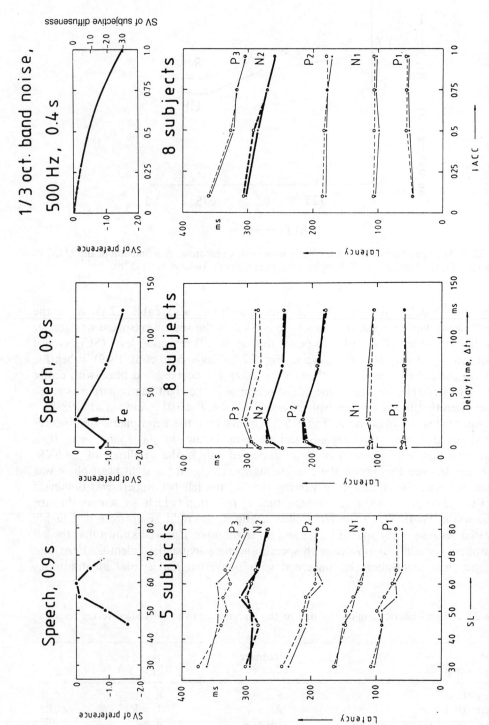

Fig. 5.2.5. Relationships between latencies of slow vertex responses and subjective preference for three objective parameters. Solid curves, left hemisphere; broken curves, right hemisphere. From Ando (1992).

(Sperry, 1974). It seems, however, that hemispheric dominance in our experiment is a relative phenomenon depending on whether temporal or spatial factors were changed in the stimulus pair: no absolute dominance was observed.

5.2.3. Relationship between N_2-Latency and Subjective Preference

Figure 5.2.5 shows the relationship between the scale values of the subjective preference and the latency components of SVR. The amplitude component indicated hemispheric dominance as discussed above. Using the paired-stimuli method, both SVR and the subjective preference for sound fields were investigated as a function of the sensation level and the time delay of the single reflection. As in the experiment described above, the source signal was continuous speech with a duration of 0.9 s. The upper parts of this figure show the results for the scale value of subjective preference, and the lower parts show the results for the latency components. As shown in the left-hand and centre columns, when the SL and the delay time of the reflection were changed, the information related to subjective preference (primitive response) typically appeared with difference of N_2-latencies in the two hemispheres in response to a pair of sound fields.

The right-hand column shows the effects of the IACC with the 1/3 octave band noise (500 Hz). In the upper part, the scale value of the subjective diffuseness is indicated as a function of IACC. The scale value of the subjective preference shows a similar response to changes in IACC when speech or music signals are presented as described in the previous section. The information related to subjective diffuseness therefore appears at N_2-latencies ranging from 260 ms to 310 ms, where a tendency of increasing latency with decreasing IACC was observed for eight subjects (except only for the left hemisphere of one subject). This relationship between IACC and the N_2-latency was linear and the correlation coefficient between them was -0.99 ($p < 0.01$).

Let us also look at the early latencies of P_1 and N_1. These were almost constant when the delay time and the IACC were changed, but information related to the sensation level or loudness was typically seen in the N_1-latency. This tendency agrees well with the results reported by Botte et al. (1975). Consequently, from 40 to 170 ms of SVR, the amplitude component shows a hemispheric dominance that indicates specialization of the left and right hemispheres. The latency differences corresponding to the sensation level are seen in the range of 120–170 ms. Finally it can be seen that the N_2-latency components in the delay range between 200 ms and 310 ms may well correspond to the subjective preference for the listening level, the time delay of the reflection, and— indirectly—the IACC. Since the longest latency was always observed at the most preferred condition, it appears that most of the brain may be relaxed at the preferred condition.

An activity after a short delay (less than 10 ms after the sound signal arrives at the eardrums) may indicate a possible mechanism for the effect of the IACC as described below.

5.3. Auditory Brainstem Responses (ABR) as a Function of the Horizontal Angle to a Listener

In the judgement of subjective preference and subjective diffuseness, a possible mechanism for the effects of IACC has been assumed in the auditory pathways. To find a mechanism by which spatial information might act in an auditory pathway, the left and right auditory brainstem responses (ABR) were recorded.

5.3.1. ABR Recording and the Flow of Neural Signals

As a source signal $p(t)$, a short-pulse signal (50 μs) was supplied to a loudspeaker with frequency characteristics ± 3 dB for 100 Hz to 10 kHz. This signal was repeated at every 100 ms for 200 s (2000 times), and left and right ABRs were recorded through electrodes placed on the vertex, and on the left and right mastoids. The distance between the loudspeaker and the centre of head was 68 ± 1 cm. The loudspeaker was located at the right-hand side of the listener. Examples of the ABR recorded from one of four subjects are shown in Fig. 5.3.1 as a parameter of the horizontal angle of sound incidence. The waves I–VI from the vertex and the right mastoid clearly differ in amplitude. Similar ABR data were obtained from four subjects (23 ± 2 years of age, male) and were averaged. As shown in Fig. 5.3.2a (wave I), what is of particular interest is that amplitudes from the right, which may correspond to the sound pressure from the source located on the right-hand side, are greater than those from the left: $R > L$ for $\xi = 30–150°$ ($p < 0.01$). (Note that there is no significant difference in the amplitude of waves I–VI at $\xi = 0°$ and 180°.) As shown in Fig. 5.3.2b, this tendency is reversed in wave II: $L > R$ for $\xi = 60°$ and 90° ($p < 0.05$). The behaviour of wave III, shown in

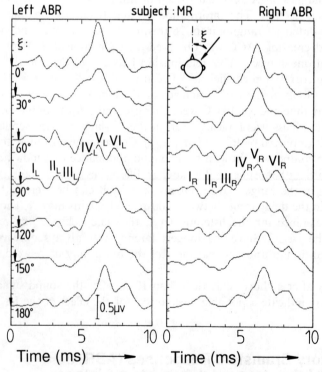

Fig. 5.3.1. Example of auditory brainstem responses recorded between the vertex and the left and right mastoids as a parameter of the horizontal angle of sound incidence. The abscissa is the time referred to the time when the single pulse arrives at the entrance of the right ear. Arrows indicate the time delay depending upon the sound source location (on the right-hand side of the subject) and the null amplitude of ABR. From Ando et al. (1991).

Fig. 5.3.2c, is similar to that of wave I: $R > L$ for $\xi = 30–150°$ $(p < 0.01)$. As shown in Fig. 5.3.2d, this tendency is reversed again in wave IV: $L > R$ for $\xi = 60°$ and $90°$ $(p < 0.05)$, and this is maintained further in wave VI as shown in Fig. 5.3.2f, even though the absolute values are greater than those of waves III: $L > R$ for $\xi = 60°$ and $90°$ $(p < 0.05)$. As shown in Fig. 5.3.3, this indicates that the flow of neural signals is probably interchanged three times between the cochlear nucleus, the superior olivary

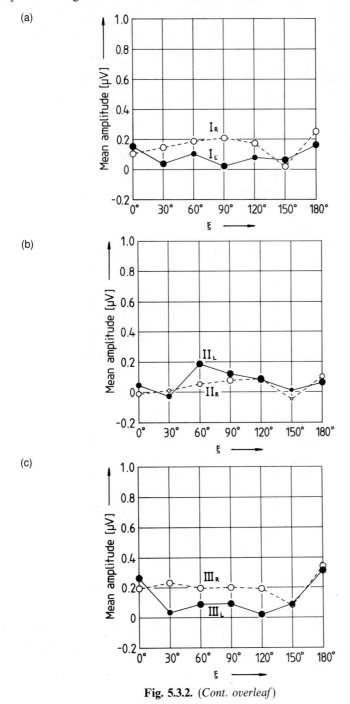

Fig. 5.3.2. (*Cont. overleaf*)

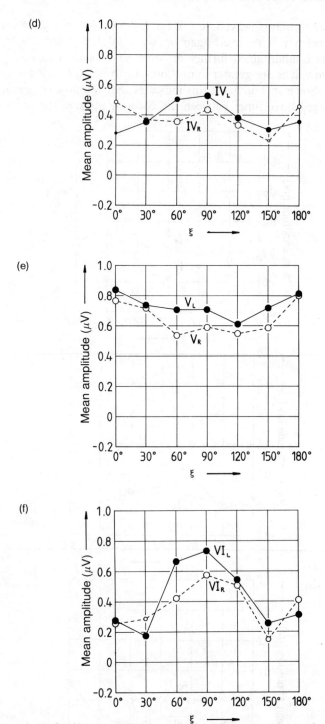

Fig. 5.3.2. Averaged amplitudes of auditory brainstem responses (ABRs) for four subjects, waves I–VI. The sizes of the circles indicate the number of available data from four subjects. Solid circles, left ABRs; open circles, right ABRs. (a) Wave I; (b) wave II; (c) wave III; (d) wave IV; (e) wave V; (f) wave VI. From Ando et al. (1991).

ABR WAVES

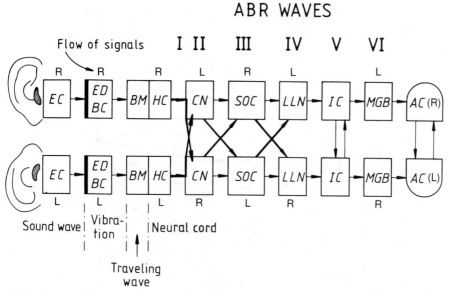

Fig. 5.3.3. Schematic illustration of the flow of signals in auditory pathways. EC, external canal; ED, eardrum; BC, born chain; BM, basilar membrane; HC, hair cell; CN, cochlear nucleus; SOC, superior olivary complex; LLN, lateral lemniscus nucleus; IC, inferior colliculus; MGB, medical denticulate body; AC, auditory cortex of the right and left hemispheres. The source location of each ABR wave has been discussed previously on the basis of recordings in animal and human subjects (Jewett, 1970; Lev and Sohmer, 1972; Buchwald and Huang, 1975).

complex and the lateral lemniscus. As discussed below, the interchanges at the inferior colliculus may be operative for the interaural signal processing.

Such a reversion cannot be seen for wave V (Fig. 5.3.2e), and the relative behaviour of amplitudes is parallel and similar on the left and the right. Thus these two amplitudes were averaged (see Fig. 5.3.6, symbols V). In Fig. 5.3.6, the amplitudes of wave IV (left and right, symbols L and R) are also plotted in reference to the ABR amplitudes at the frontal sound incidence.

Concerning the latencies of waves I through VI in reference to the time when the short pulse was supplied to the loudspeaker, the behaviours indicating relatively short latencies in the range around $\xi = 90°$ were similar (Fig. 5.3.4). It is remarkable that the significant ($p < 0.01$) difference between the averaged latencies at $\xi = 90°$ and those at $\xi = 0°$ and $180°$, is about $640 \, \mu s$, which is approximately the same with the interaural time difference of sound incidence at $\xi = 90°$. It is most likely that the relative latency of wave III typically may be reflected by the interaural time difference. As indicated in Fig. 5.3.4, there was no significant left–right differences between the latencies of waves I–IV.

5.3.2. ABR Amplitudes in Relation to IACC

The magnitudes of interaural cross correlation and autocorrelation functions at $\tau = 0$ of the signals after passing through the A-weighting networks, measured at the two ear

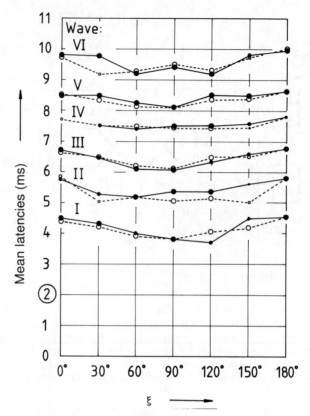

Fig. 5.3.4. Averaged latencies of the auditory brainstem responses (ABRs) for four subjects, waves I–VI. Sizes of the circles indicate the number of available data. The latency of 2 ms indicated by ② corresponds to the distance between the loudspeaker and the centre of the head. Solid circles, left ABRs; open circles, right ABRs. From Ando et al. (1991).

entrances of a dummy head, are shown in Fig. 5.3.5 as a function of the horizontal angle of the sound incidence. The averaged amplitudes of wave IV (left and right) and of wave V, both normalized to their amplitudes at the frontal incidence ($\xi = 0$), are shown in Fig. 5.3.6. Similar results were obtained when amplitudes were normalized by those at $\xi = 180°$. Although the results shown in Figs. 5.3.5 and 5.3.6 cannot be compared directly, it is interesting to point out that the relative behaviour of wave IV (L) in Fig. 5.3.6 is similar to $\Phi_{RR}(0)$ measured at the right-ear entrance. And the relative behaviour of wave IV (R) is similar to the $\Phi_{LL}(0)$ measured at the left-ear entrance. In fact, the amplitudes of wave IV (left and right) are proportional to $\Phi_{xx}(0)$, $x = L, R$. The behaviour of wave V is similar to the maximum value, $|\Phi_{LR}(\tau)|_{max}$, $|\tau| < 1$ ms. Since correlations have the dimension of the power of sound signals (i.e. since they have an order of A^2), the IACC defined by Eq. (5.1.8) may correspond to

$$P = \frac{A_V^2}{[A_{IV,R} A_{IV,L}]} \tag{5.3.1}$$

where $A_{IV,R}$ and $A_{IV,L}$ are the respective amplitudes of wave IV on the right and left, and where A_V is the amplitude of wave V which may be reflected by the 'maximum'

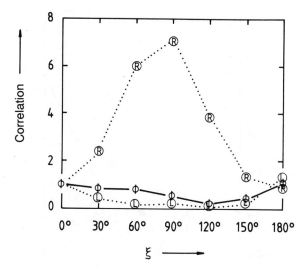

Fig. 5.3.5. Maximum correlations of sound signals at the left-ear and right-ear entrances of a dummy head. ®, ACF at $\tau = 0$, $\Phi_{RR}(0)$ measured at the right-ear entrance. ⓛ, ACF at $\tau = 0$, $\Phi_{LL}(0)$ measured at the left-ear entrance. ϕ, maximum ICF, $|\Phi_{LR}(\tau)|_{max}$, $|\tau| < 1$ ms. From Ando et al. (1991).

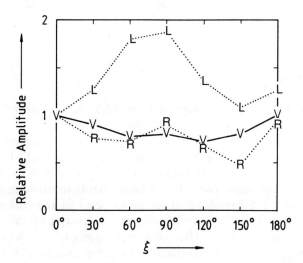

Fig. 5.3.6. Averaged amplitudes of wave IV (left and right) and averaged amplitudes of wave V normalized to the amplitudes at the frontal incidence (4 subjects). From Ando et al. (1991).

neural activity $[A_V^2 \approx |\Phi_{LR}(\tau)|_{max}]$ at the inferior colliculus with signal interchanges of the left and right (see Fig. 5.3.3).

The results obtained by using Eq. (5.3.1) are plotted in Fig. 5.3.7. It is clear that relative behaviours of IACC and P are in good agreement, except for the value P at $\xi = 150°$, where only a single data point from a single subject was obtained for A_{IV}. The correlation coefficient between the values of IACC and P is 0.92 ($p < 0.005$).

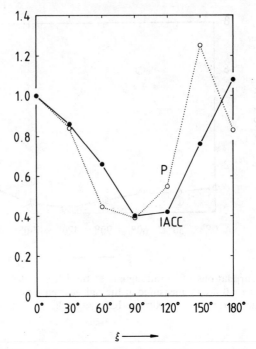

Fig. 5.3.7. Values of IACC and P calculated by Eqs. (5.1.7) and (5.3.1), respectively. IACC > 1 is due to errors of measured values. From Ando et al. (1991).

5.3.3. Remarks

As shown in Fig. 5.3.2, the amplitudes of ABR clearly depends on the horizontal incidence angle. In particular, the amplitudes of waves IV_L and IV_R are nearly proportional to the sound pressures at the right and left ear entrances, respectively, when the amplitude is normalized to that at $\xi = 0°$ or 180°. Ino et al. (1985) also reported that the auditory electromotive vector changed with the horizontal angle.

As far as the amplitudes of the left and right ABRs recorded here are concerned, the first interchange of a neural signal seems to occur at the entrance to the cochlear nucleus, the second at the superior olivary complex, and the third at the lateral lemniscus nucleus (Fig. 5.3.3). Thompson and Thompson (1988a, b), who traced these neuroanatomical tracts in guinea-pigs, found four separate pathways, all via brainstem neurons connecting one cochlea either with the other cochlea or with itself. Those pathways may be related to the first interchange at the entrance of cochlea. The 'maximum' neural activity may correspond to the IACC appearing in the amplitude of ABR (wave V) about 8 ms after the sound signal supplied to loudspeakers, which includes the sound propagation time from the listener (68 ± 1 cm) and the neural delay. Also, the relative latency of wave III corresponds to the interaural time difference (Fig. 5.3.4).

It should be noted here that in recording ABRs, a short-pulse signal like that applied here (50 μs) should be supplied to a loudspeaker with a wide frequency range because the interval between ABR waves is of the order of 1 ms. And the use of headphones

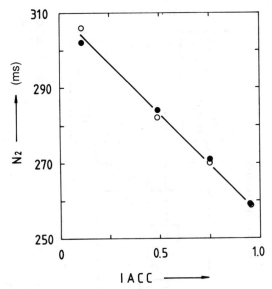

Fig. 5.3.8. N_2-latency of slow vertex response (SVR) as a function of the IACC (see also Fig. 5.2.5). Solid circles, N_2-latency of SVR over the left hemisphere; open circles, N_2-latency of SVR over the right hemisphere; solid line, regression line. From Ando et al. (1987b).

may cause additional uncertainty in this kind of investigation because listening conditions are unusual.

From the results of the previous investigation shown in Fig. 5.3.8 (see also Fig. 5.2.5 right), it is remarkable that a linear relationship between the IACC and the N_2-latency is observed in the slow vertex response of both cerebral hemispheres ($p < 0.01$). Thus, the subjective preference and subjective diffuseness judgements of the sound field previously described in relation to the IACC are well based on neural activity in the auditory–brain system.

5.4. Model of the Auditory–Brain System

The above subjective and physiological responses can be used as the basis for a model of the auditory–brain system. Major independent acoustic factors classified by comprehensive temporal and spatial factors and represented in the model can be used in designing sound fields. The model consists of the auto correlation mechanisms, the mechanism of interaural cross correlation between the two auditory pathways, and the specialization of human cerebral hemispheres for processing temporal and spatial factors of a sound field.

5.4.1. Model of the Auditory–Brain System

The model shown in Fig. 5.4.1 is based on the subjective attributes and on the auditory potentials evoked in response to the change of acoustic factors. The sound source $p(t)$ in this figure, is located at r_0 in a 3-dimensional space and a listener is sitting at r

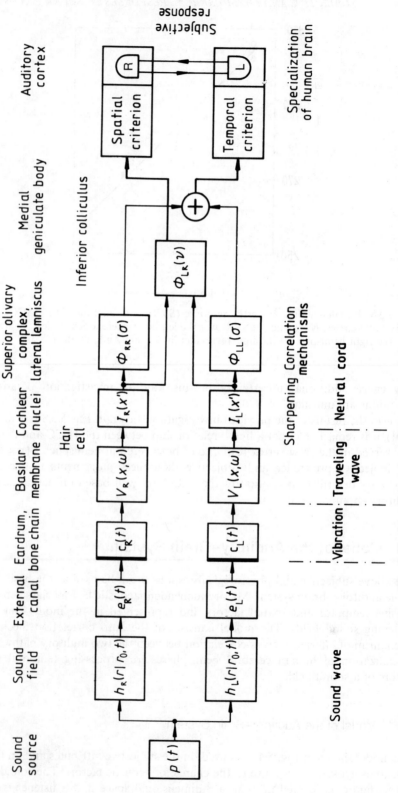

Fig. 5.4.1. A model of the auditory–brain system. From Ando (1985).

(defined by the location of the centre of the head), $h_{L,R}(r\,|\,r_0, t)$ being the impulse responses between r_0 and the left and right ear-canal entrances. The impulse response of the external ear canal is $e_{L,R}(t)$ and that of the bone chain is $c_{L,R}(t)$. The velocities of the basilar membrane are expressed by $V_{L,R}(x, \omega)$, x being the position along the membrane. As shown in Fig. 5.4.1, the action potentials from the hair cells are conducted to the cochlear nuclei, the superior olivary complex (including the medial superior olive, the lateral superior olive, and the trapezoid body), and the higher levels of the cerebral hemispheres (see Fig. 5.3.3 also).

According to the tuning of a single fibre (Katsuki et al., 1958), the input spectrum of the cochlea $I(x')$ can be roughly mapped at a certain nerve position x'. This mapping may be partially supported by the ABR waves (I–IV), which reflect the sound pressure levels as a function of the horizontal angle of incidence (Section 5.3). Such neural activities, in turn, include enough information to generate the ACF at a higher level, probably near the lateral lemniscus as indicated by $\Phi_{LL}(\sigma)$ and $\Phi_{RR}(\sigma)$. For convenience, the interchange of neural signals discussed in Section 5.3 is not considered here.

Also as discussed in Section 5.3, the neural activity (wave V) seems to correspond to the IACC, and thus the interaural cross correlation mechanism may exist at the inferior colliculus. We described in Section 5.2 that the sound pressure is processed by the ACF at the time of origin ($\sigma = 0$) and that the output signal of the interaural cross correlation mechanism, including the IACC and the loci and the sharpness of maxima, may be predominantly connected to the right hemisphere. Effects of the initial time delay between the direct sound and the single reflection Δt_1 included in the autocorrelation function ($\sigma \neq 0$) may activate the left hemisphere. The specialization of the human cerebral hemispheres may relate to the highly independent contributions of spatial and temporal factors to the subjective attributes. Appropriate data about the reverberation time are not yet available, but this effect may occur in the left hemisphere because of the temporal factor being processed by autocorrelators. It is remarkable that 'cocktail party effects', for example, may be explained by lateral specialization of the human brain, since spatial information is processed mainly in the right hemisphere. It is worth noting here that speech is predominantly processed in the left hemisphere and the spatial information is predominantly processed in the right hemisphere (Sperry, 1974).

5.4.2. Review of Subjective Responses

So-called 'overall responses' like subjective preferences must typically be associated with both hemispheres and with all the acoustic factors of the temporal and spatial variables. It is interesting that speech intelligibility and clarity may be influenced by all of the four factors, including the IACC (Nakajima and Ando, 1991; Nakajima, 1992), associated with both hemispheres. In general, therefore, any subjective attributes are more or less associated with both cerebral hemispheres. It is also worth noting that subjective diffuseness is considered to depend mainly on the IACC and the listening level (Keet, 1968). These factors may be associated with activity in the right hemisphere (Table 5.2.1), but there is no evidence that subjective diffuseness is independent of the reverberation time. Furthermore, although the echo disturbance, the perception threshold of reflection and coloration with temporal information are considered to be predominantly processed in the left hemisphere and to be related to the ACF of the source signals and Δt_1 (Ando, 1986), the echo disturbance may depend more or less on the IACC (Fig. 5.6.5). In addition, the IACC as a subjective diffuseness factor (Ando and Kurihara, 1986; Singh, Ando and Kurihara, 1994) is independent of Δt_1, except for the increase in IACC occurring in the very short delay range $0 < \Delta t_1 < 0.1\tau_e$.

   In such a manner, any other subjective responses may be well described by the four orthogonal factors together with the ACF of source signals and the total amplitude of reflections. Further investigations, however, are necessary to identify some other subjective attributes.

5.5. Individual Preference

Individual preference may be described according to the theory derived from 'global' subjective preference measured with a number of subjects as well as according to the model of the auditory–brain system described above (Singh and Ando, unpublished). The limited number of orthogonal acoustic factors for the sound field in a room may be used to identify the most preferred condition for each individual listener, and such individual characteristics may be utilized in selecting a seat in a concert hall.

5.5.1. Analysis of Individual Preference

A theory similar to the global preference given by Eqs (5.1.12) and (5.1.13) may also be derived for each individual level (Singh and Ando, unpublished). First of all, as discussed in previous sections, the independence of acoustic factors may be described in terms of the lateral specialization of the human cerebral hemispheres. According to the amplitude of the early SVR, the listening level (sensation level) and the IACC seem to be right-hemisphere dominant (spatial orientation), and the initial time delay Δt_1 seems to be left-hemisphere dominant (temporal orientation). The reverberation time is not identified by the analysis of brain waves, but because its temporal factor is similar to that of Δt_1, it is assumed to be closely associated with the left hemisphere. Thus, it is considered that right-hemisphere spatial factors and left-hemisphere temporal factors independently influence the subjective preference of each individual.

   Next, to demonstrate effects of both spatial and temporal criteria upon the scale value of subjective preference judgements, an approximate method of analysing individual data is proposed. On the basis of the law of comparative judgement, case V (Thurstone, 1927), the global scale value S_s of sound fields is given by

$$S_s = \frac{1}{N} \sum_{j=1}^{N} Z_{sj}, \qquad s, j = 1, 2, \ldots, N \qquad (5.5.1)$$

where Z_{sj} is obtained by the probability that the sound field s is preferred to sound field j, such that

$$P_{sj} = P(s > j) = \frac{1}{\sqrt{2\pi}} \int_{Z_{js}}^{\infty} \exp\left(-\frac{y^2}{2}\right) dy \tag{5.5.2}$$

thus $Z_{js} = L(P_{sj})$.

The probability, obtained by the number of positive judgements Y_{js}, is given by

$$P_{js} = \frac{1}{M} \sum_{h=1}^{M} Y_{js,h}, \qquad h = 1, 2, \ldots, M \tag{5.5.3}$$

where M is the number of individuals. Therefore,

$$S_s = \frac{1}{N} \sum_{j=1}^{N} L(P_{sj})$$

$$= \frac{1}{N} \sum_{j=1}^{N} L\left(\frac{1}{M} \sum_{h=1}^{M} Y_{js,h}\right) \tag{5.5.4}$$

Over a linear range between P_{sj} and Z_{sj}, or $0.05 < P_{js} < 0.95$, Eq. (5.5.4) may be expressed by change of summation order:

$$S_s \approx \frac{L_0}{MN} \sum_{h=1}^{M} \sum_{j=1}^{N} Y_{js,h}$$

$$= \frac{L_0}{M} \sum_{h=1}^{M} P_{s,h} \approx \frac{1}{M} \sum_{h=1}^{M} Z_{s,h} \tag{5.5.5}$$

where L_0 is a linear operator and the probability $P_{s,h}$ is obtained by the number of positive judgements with sound fields N. To keep a certain accuracy, it is recommended that

$$N \gg 2 \tag{5.5.6}$$

Thus, the scale value of each individual h may be analysed as an element of the global scale value,

$$S_{s,h} \approx Z_{s,h}, h = 1, 2, \ldots, M \tag{5.5.7}$$

So far we have considered Thurstone's case V with an identical standard deviation of the discriminative process in the preference judgement between sound fields s and j, as a unit.

In this manner, global preference data (Ando and Morioka, 1981; Ando et al, 1982) were reanalysed as shown in the following section finding substantial individual differences.

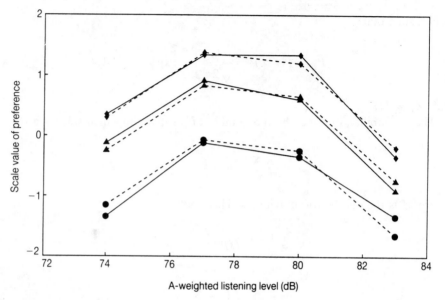

Fig. 5.5.1. An example showing the independence of listening level and IACC. Solid line, scale values obtained by the preference tests. Dotted line, scale values calculated by the superposition due to the effects of the listening level and the IACC.

5.5.2. Relationship between Individual Preference and the Four Orthogonal Factors

A. *Listening Level and IACC*

The subjective preference data obtained when changing both listening level and IACC (Ando and Morioka, 1981) were analysed again by using the methods described in the preceding subsection. Analyses of variance indicated that listening level and IACC are independent of the preference judgements for 'each listener' (16 subjects, with music motifs A and B). A typical example showing the independence of the two factors in a single subject (OS) is shown in Fig. 5.5.1. Despite the right-hemispheric localization of these factors, no interaction is seen between the parallel curves for the IACC and for the listening level. The scale values of the two factors are thus described separately and superposed in a similar manner using Eq. (5.1.12).

Individual scale values of preference with music motif A are shown for 16 subjects in Figs 5.5.2 and 5.5.3 as functions of listening level and of IACC. Large individual differences exceeding the range 74–83 dBA are found in the most preferred listening level. Subjects OK, AK, and YU, for example, preferred levels below A-weighted 74 dB, whereas subjects MA and HY preferred levels above A-weighted 83 dB.

As shown in Fig. 5.5.3, on the other hand, all 16 subjects indicated higher preference with decreasing IACC regardless of the music motif used in the tests. It is remarkable that, as indicated in Fig. 5.5.4, the most preferred listening level of each individual for music motif A is correlated with that for music motif B (the correlation coefficient is above 0.6).

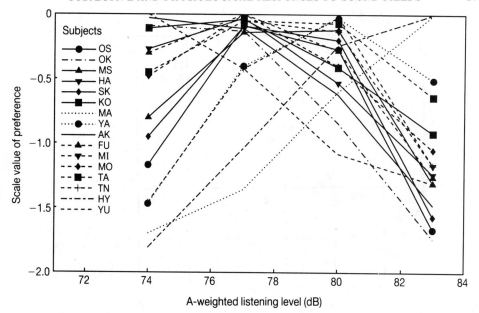

Fig. 5.5.2. Scale values of individual preference as a function of the listening level (16 subjects, music motif A).

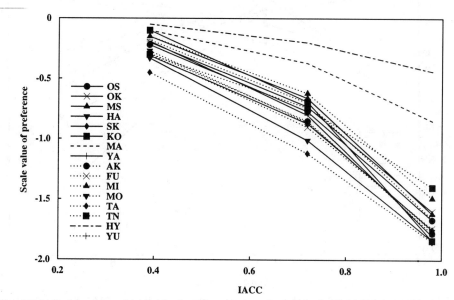

Fig. 5.5.3. Scale values of individual preference as a function of the IACC (16 subjects, music motif A).

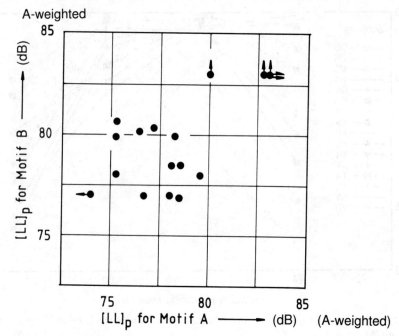

Fig. 5.5.4. The most preferred listening levels $[LL]_p$ of each listener with music motifs A and B.

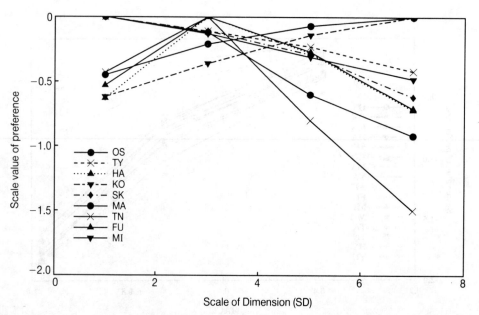

Fig. 5.5.5. Scale values of individual preference as a function of SD ($\approx 22\Delta t_1$) (9 subjects, music motif A).

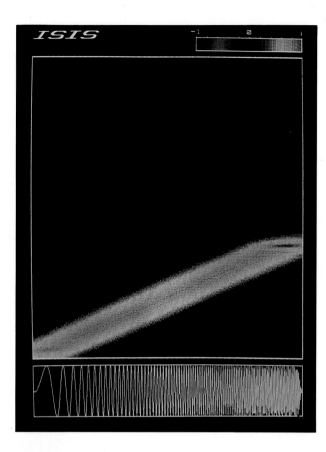

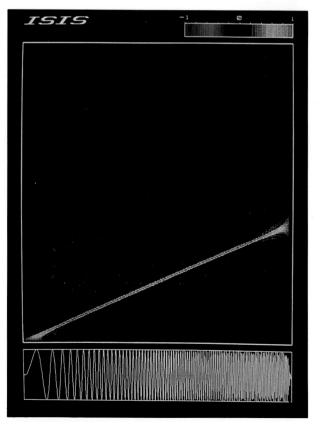

Plate I. Time–frequency representations of a chirp signal by spectrogram (top) and Wigner distribution (left).

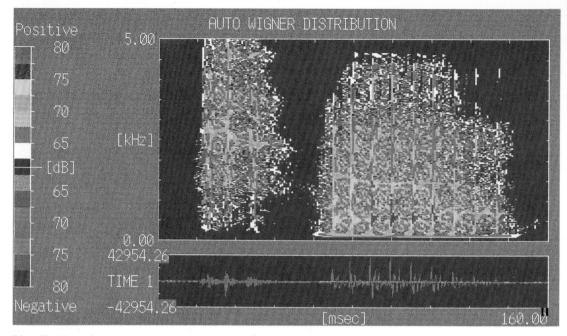

Plate II. Time–frequency representation of a vocalization ('cat') by the Wigner distribution.

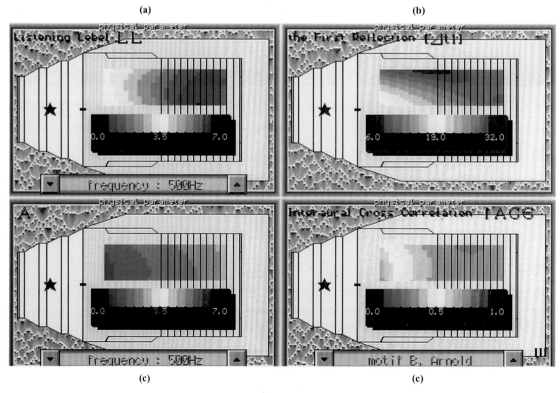

Plate III. Calculated value of physical factors at each seat. (a) Listening level (500 Hz). (b) Initial time delay gap. (c) Total amplitude of reflection, A. (d) IACC (music motif B).

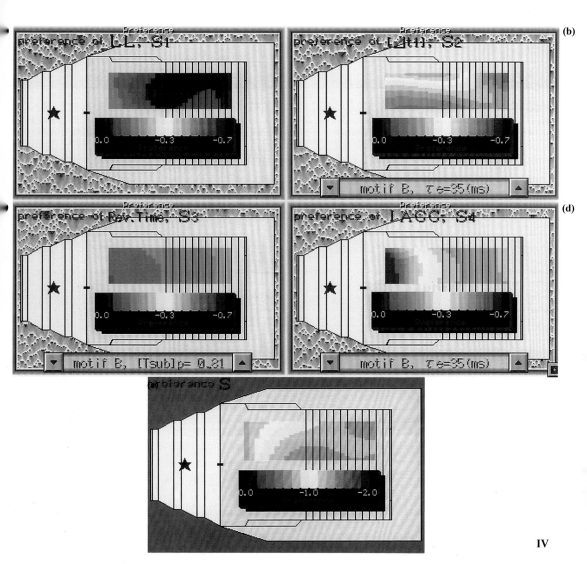

Plate IV. Calculated scale value of subjective preference at each seat. (a) Scale value of preference S_1. (b) Scale value of preference S_2. (c) Scale value of preference S_3. (d) Scale value of preference S_4. (e) Total scale value of preference, $S = S_1 + S_2 + S_3 + S_4$.

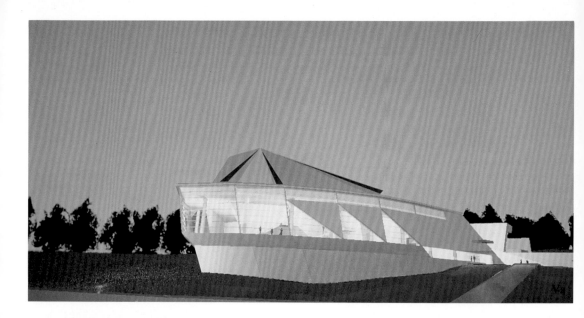

Plate V. Kirishima International Concert Hall designed by Maki and Associates and for which a simulation system for testing the subjective preference of a sound field by all listeners was planned, and realized in 1994.

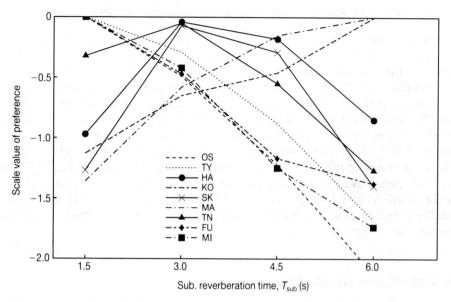

Fig. 5.5.6. Scale values of individual preference as a function of the subsequent reverberation time (9 subjects, music motif A).

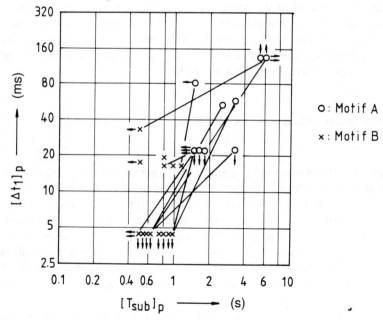

Fig. 5.5.7. Preference distribution of the most preferred reverberation time T_{sub} and Δt_1 (music motifs A and B).

B. *Delay Time of Early Reflections and Subsequent Reverberation Time*

Since the delay time of early reflections and the subsequent reverberation time influence the subjective preference scale for each listener independently, these effects can also be described separately. Individual scale value of preferences are shown in Figs 5.5.5 and 5.5.6 as functions of SD (scale of dimension of concert halls, $\Delta t_1 \approx 22$ SD (ms)) and of T_{sub}. Both the preferred initial time delay (SD or Δt_1) and the preferred T_{sub} differed greatly between individuals. As shown in Fig. 5.5.7, these most preferred conditions are somewhat related to each other because both are obtained in relation to the effective duration of ACF of the source signal ($\tau_e \approx 127$ ms for motif A and $\tau_e \approx 43$ ms for motif B). It is worth noting that the most preferred conditions obtained by Eqs (5.1.9) and (5.1.11) for global listeners (16 subjects) are $[\Delta t_1]_p \approx 55$ ms ($A = 3.7$) and $[T_{sub}]_p \approx 2.9$ s for motif A, and $[\Delta t_1]_p \approx 14.5$ ms ($A = 4.6$) and $[T_{sub}]_p \approx 0.99$ s for motif B, respectively. But Δt_1 and T_{sub} are independent as far as the scale values of individual preference are concerned.

5.6. Subjective Attributes

The subjective preference and related neural activities discussed in the preceding section can be used to describe other important subjective attributes in relation to the ACF of source signals and to the temporal factors and the spatial factor (IACC) of a sound field. Loudness, for example, one of the most fundamental subjective responses, may change according to a repetitive feature of the source signal (ACF, τ_e) as well as according to the level of the sound pressure. The threshold of perception of a single reflection and the just-noticeable difference appearing as coloration by a single reflection may be well expressed by the ACF envelope of the primary sound together with its delay time. The subjective diffuseness of a sound field may be related to the IACC when other orthogonal factors are fixed.

5.6.1. Subjective Attributes Due to the ACF of the Source Signal

A. *Loudness*

With the fixed level of sound pressure (74 dB) and other temporal and spatial factors, the loudness was judged in response to change in the ACF of bandpass noise within the critical band (Merthayasa and Ando, unpublished). The effective duration of ACF, τ_e, of the bandpass noise of the 1 kHz centre frequency was controlled by a bandpass filter with ± 48, ± 140 and ± 1080 dB/octave slopes. The duration of stimuli was 3 s, their rise and fall time was 250 ms, and the interval between stimuli was 1 s. The bandwidth ΔF defined by the -3 dB attenuation of the low and high cut-off frequencies of the stimuli was kept constant at a certain value. In fact, ΔF was set tending to '0 Hz' as well to control a wide range of τ_e (3.5–52.6 ms).

The paired-comparison method for judgements was used for six male students with normal hearing ability. Because the subjects sat in a semi-anechoic chamber facing a loudspeaker 90 ± 1 cm in front of them, the IACC was kept constant at nearly unity. The scale values of loudness are shown in Fig. 5.6.1 as a function of τ_e. Statistical analysis with different values of τ_e indicates significant differences ($p < 0.01$)

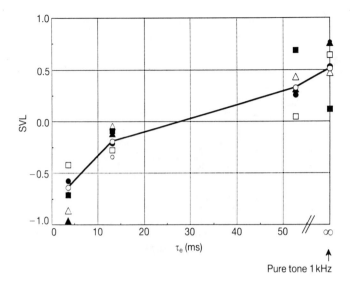

Fig. 5.6.1. Scale value of loudness (SVL) as a function of τ_e. Different symbols indicate the scale values obtained with 6 different subjects.

in loudness except between the pure tone stimuli and the stimuli with $\tau_e = 52.6$ ms. Obviously, the loudness is influenced by increasing the value of τ_e, and this may demonstrate that the degree of the repetitive feature of stimuli contributes to the loudness. Since there was no significant loudness difference between the pure tone and the '0 Hz' bandwidth signal produced using filters with 1080 dB/octave slopes, it is recommended that a filter with such a sharp slope be used in every hearing experiment that depends on the sharpness of a filter in the auditory system.

As shown in Fig. 5.6.2, when a filter slope of 1080 dB/octave was used, the loudness of bandpass noise within the critical band was not constant but had a minimum at a certain bandwidth. This differs from the results by Zwicker et al. (1957), except for the pure tone, which was indicated as being louder than the other stimuli. It is worth noting that the loudness does not depend on the IACC when the level of sound pressure at both ear entrances is fixed (Ando and Nagatani, unpublished). This confirms results obtained using headphone reproduction (Dubroskii and Chernyak, 1969).

B. *Subjective Preference*

Results obtained using subjective preference as an overall psychological response to a sound field with a single reflection indicated that the most preferred delay of the reflection may be found within the envelope curve of the ACF defined by the delay (Ando, 1977):

$$[\Delta t_1]_p = \tau_p$$

such that

$$|\phi(\tau)|_{\text{envelope}} \approx k A_1^c \qquad \text{at } \tau = \tau_p \tag{5.6.1}$$

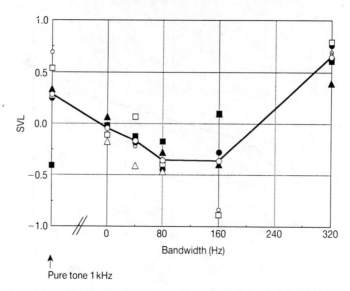

Fig. 5.6.2. Scale values of loudness (SVL) as a function of the bandwidth. Different symbols indicate the scale values obtained with 6 different subjects.

where A_1 is the pressure amplitude of the single reflection in reference to that of the direct sound, $k = 0.1$, and $c = 1$. If the envelope of the ACF is exponential, this equation can be simply expressed by Eq. (5.1.9).

Such a relationship may also be found for other subjective responses in relation to the temporal factor. Constants related to the ACF envelope k, c, and A_1 used in calculating various subjective responses to sound fields, with limited ranges of experiments are listed in Table 5.6.1.

Table 5.6.1. Constants related to the ACF envelope and used for calculating various subjective responses to sound fields with single reflection

Subjective response	(in Eq. (5.6.1)) k	c	Delay time to be obtained	Range of amplitude examined (dB)	Source signal
Preference of listeners	0.1	1	Preferred delay time	$7.5 \geqslant A_1 \geqslant -7.5$	Speech and music
Threshold of perception of reflection	2	1	Critical delay time	$-10.0 \geqslant A_1 \geqslant -50.0$	Speech
50% echo disturbance	0.01	4	Disturbed delay time	$0 \geqslant A_1 \geqslant -6.0$	Speech
Coloration	$10^{-2.5}$	-2	Critical delay time	$-7.0 \geqslant A_1 \geqslant -27.0$	Gaussian noise
Preference of musicians	2/3	1/4	Preferred delay time	$-10.0 \geqslant A_1 \geqslant -34.0$	Music

C. *Coloration*

Coloration is clearly perceived if we listen to sound when we are very near to a boundary wall in a room. Such coloration is discussed here in relation to the ACF envelope. Continuous bandpass noise was used as a source signal because the ACF of the signal is independent of the time interval extracted for subjective judgements and is available theoretically. After Gaussian noise passes through an ideal bandpass filter with a flat response between frequencies f_1 and f_2, $f_2 > f_1$, its normalized ACF may be written as

$$\phi(\tau) = \frac{2}{\Delta\omega\tau} \sin\left(\frac{\Delta\omega\tau}{2}\right) \cos\left(\frac{\Delta\omega_c\tau}{2}\right) \tag{5.6.2}$$

where

$$\Delta\omega = 2\pi(f_2 - f_1)$$

and

$$\Delta\omega_c = 2\pi(f_2 + f_1)$$

The envelope of the ACF is expressed by

$$\frac{2}{\Delta\omega\tau} \sin\frac{\Delta\omega\tau}{2}, \qquad \text{for } 0 \leqslant \Delta\omega\tau \leqslant \pi$$

and by

$$\frac{2}{\Delta\omega\tau}, \qquad \text{for } \Delta\omega\tau > \pi \tag{5.6.3}$$

Figure 5.6.3 shows an example of the measured ACF for the bandpass noise reproduced by a loudspeaker in an anechoic chamber, and also shows the envelope curve calculated by putting an equivalent bandwidth for $\Delta\omega$ into Eq. (5.6.3). Since the nonideal characteristics of the filter and the loudspeaker were used, the equivalent bandwidth had to be chosen greater than the difference of the cut-off frequencies in Eq. (5.6.2). The loudspeaker arrangement presenting the primary sound and the delayed weak sound is shown in the right-hand part of Fig. 5.6.4. The total pressure of the two sounds was automatically kept constant (A-weighted 60 dB) at the listener's position in the anechoic chamber. The subject adjusted the level of sound pressure for the weak delayed sound and judged the threshold of a just-noticeable difference appearing as a coloration in comparison with the situation in which only the primary sound was presented.

For the noise source with a centre frequency of 1 kHz, the threshold levels of the weak sound relative to the primary sound are shown in Fig. 5.6.4 as a function of the delay time Δt_1. The dashed curve shows the values calculated from Eqs (5.6.1) and (5.6.3) with the ACF envelope (Fig. 5.6.3) and the derived constants: $k = 10^{-2.5}$ and $c = -2.0$. Similar results were obtained with 250 Hz and 4 kHz even when the direction of the weak sound, $\xi = 36°$, $\xi = 90°$, was changed (Ando and Alrutz, 1982).

D. *Threshold of Perception of a Single Reflection*

Seraphim (1961) investigated the absolute perceptible threshold (aWs) of a single reflection with speech sound (Fig. 5.6.5). Unfortunately, the ACF of the speech signal

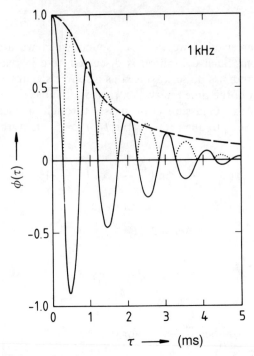

Fig. 5.6.3. Measured ACF of the bandpass noise with a centre frequency of 1 kHz. Solid curve, measured ACF; dotted curve, absolute values of the measured ACF; broken curve, envelope curve by Eq. (5.6.3). From Ando et al. (1982).

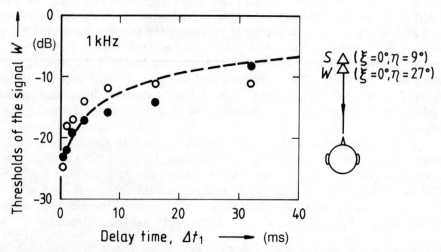

Fig. 5.6.4. Threshold level of the delayed weak sound W as a function of the delay time Δt_1. Different symbols indicate the response of two subjects. Broken curve, values calculated by using Eqs (5.6.1) and (5.6.3) with $k = 10^{-2.5}$ and $c = -2.0$. From Ando et al. (1982).

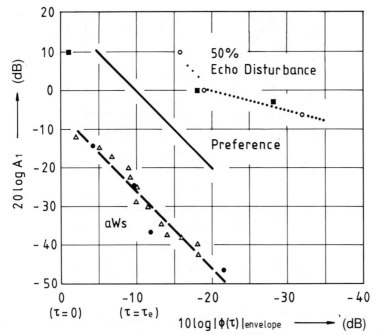

Fig. 5.6.5. Amplitude of the single reflection (obtained at several subjective responses) in relation to the ACF envelope of source signals. Solid curve, most preferred condition for listener; broken curve, threshold of perception of the reflection; dotted curve, 50% echo disturbance; triangles, aWs by Beurtleilungsverfahre (Seraphim, 1961) arranged with the ACF envelope shown in Fig. 5.6.6; solid circles, threshold of perception by the method of limit (Morimoto et al., 1983) arranged with the ACF envelope shown in Fig. 5.6.6; open circles and solid squares, 50% echo disturbances, after Haas (1951) and Ando et al. (1974), with the ACF envelope shown in Fig. 5.6.6. From Ando (1986).

that was used there is not available for examination of its relation to the behaviour of the aWs. But the ACFs of any continuous speech signals at normal speed are probably little different. Let us apply a typical ACF envelope function of the speech signal as shown in Fig. 5.6.6. Surprisingly, the aWs may be described with the ACF envelope as indicated at the lower part of Fig. 5.6.5. The aWs data rearranged were obtained under the condition of a single reflection with a horizontal angle of $\xi = 30°$ to the listeners (Seraphim, 1961). To confirm this result, threshold values of perception obtained by the method of limit (Morimoto et al., 1983) are plotted in Fig. 5.6.5, where the continuous speech signal of the ACF envelope shown in Fig. 5.6.6 was used. Similar values were obtained despite different speech signals being used. Such a perception threshold may be well described by Eq. (5.6.1) with $k = 2.0$ and $c = 1.0$.

E. *Echo Disturbance*

In a manner similar to that described above, the echo disturbance data of Haas (1951) and of Ando et al. (1974), may be rearranged with the ACF envelope. The 50% echo disturbances are shown in the upper part of Fig. 5.6.5. Since echo disturbance effects are unclear when the delay of the single reflection is short, within 50 ms, only those

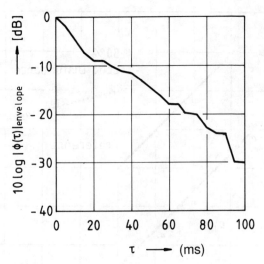

Fig. 5.6.6. Typical ACF envelope of continuous speech signals. From Ando (1986).

constants in Eq. (5.6.1) for delays longer than 50 ms ($10 \log|\phi(\tau)|_{\text{envelope}} \leqslant -20\,\text{dB}$) are meaningful for obtaining $k = 0.01$ and $c = 4.0$.

F. *Preferred Delay of a Single Reflection for a Performer*

From preference judgements with respect to the ease of performance by alto-recorder soloists (Nakayama, 1984), the most preferred delay time of the single reflection may also be described by the expression of Eq. (5.6.1). In this case, the coefficients are $k = 2/3$ and $c = 1/4$. The coefficient k for the performers differs by a factor of about 7 from that of coefficient for the listeners ($k = 0.1$). This indicates that the evaluated amplitude of the reflection is about 7 times greater than it is for listeners, thus showing a 'missing reflection for performers' or significant effects of the reflection for performers.

5.6.2. Subjective Responses in Relation to the IACC

The ICF is a significant factor determining the perceived horizontal direction of sound and the degree of subjective diffuseness for the sound field. A well-defined direction is perceived when the normalized ICF has one sharp maximum, but a low value of the IACC (< 0.15) corresponds to subjective diffuseness or to no impression of spatial direction. If the sounds arriving at both ears are dissimilar (IACC $\rightarrow 0$), then signals that are different but that contain the same information are conveyed through the two channels of the auditory system to the brain. As discussed in Section 7.1.5 as well, this may improve speech intelligibility and speech clarity.

A. *Subjective Diffuseness in Relation to the IACC*

The subjective diffuseness, or the spatial impression of a sound field in a room, is one of the important attributes describing good acoustics of a room. To obtain the scale

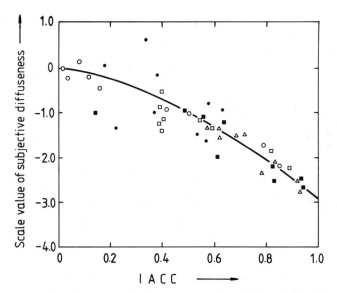

Fig. 5.6.7. Scale values of subjective diffuseness as a function of the calculated IACC. Different symbols indicate different frequencies of the bandpass noise. △, 250 Hz; ○, 500 Hz; □, 1 kHz; ●, 2 kHz; ■, 4 kHz. Solid line is regression line by Eq. (5.6.4). From Ando et al. (1986).

value of subjective diffuseness, paired-comparison tests were made using the bandpass Gaussian noise and varying the horizontal angle of two symmetric reflections (Ando and Kurihara, 1986). Listeners judged which of the two sound fields was perceived as more diffuse. A remarkable finding is that the scale values of subjective diffuseness are inversely proportional to the IACC and, like the subjective preference values (see Eq. 5.1.13) may be formulated in terms of the 3/2 power of the IACC:

$$S = -\alpha(IACC)^b, \qquad \text{where } \alpha = 2.9 \text{ and } b = 3/2 \qquad (5.6.4)$$

Results of scale values evaluated by the paired-comparison test and the values calculated by using Eq. (5.6.4) are shown in Fig. 5.6.7. There is a great variation of the data in the range IACC < 0.5, but there is no essential difference at frequencies between 250 and 4000 Hz.

The scale values of subjective diffuseness, which depend on the horizontal angle, are shown in Fig. 5.6.8 for bandpass noise with centre frequencies at 500 Hz, 1 kHz, or 2 kHz. Obviously, the most effective horizontal angles of reflection differ according to the frequency range and are inversely related to the behaviour of the IACC values (Fig. 5.6.9). These angles are about ±90° for the 500-Hz range and below, about ±55° for the 1-kHz range, and about ±20° for the 2-kHz range. For a sound field with a predominantly low-frequency range (below 250 Hz), the interaural cross correlation function has no sharp peaks when the delay $|\tau| < 1$ ms. Thus, the scale values of subjective diffuseness may increase with increasing low-frequency components.

B. *Speech Intelligibility and Clarity in Relation to IACC*

As far as speech intelligibility is concerned, Houtgast *et al.* (1980) proposed a speech transmission index (STI) that takes into consideration only the temporal and monaural

(a)

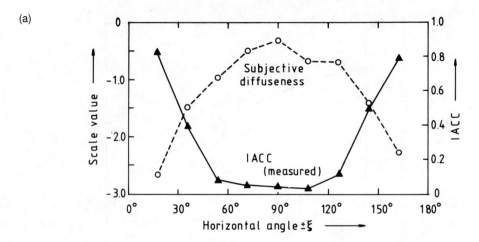

(b)

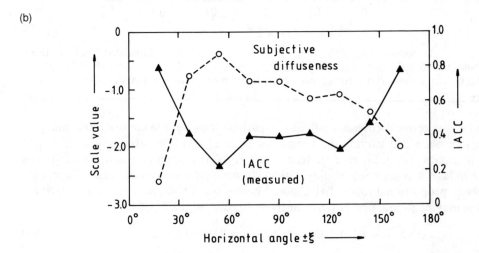

(c)

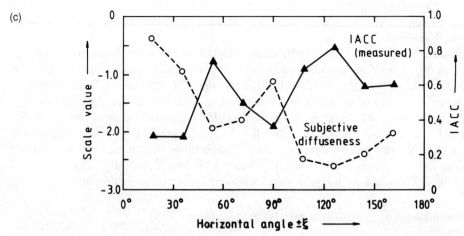

Fig. 5.6.8. Scale values of subjective diffuseness and the IACC as a function of the horizontal angle of incidence, with 1/3-octave band noise of centre frequencies: (a) 500 Hz, (b) 1 kHz, (c) 2 kHz. From Singh et al. (1994).

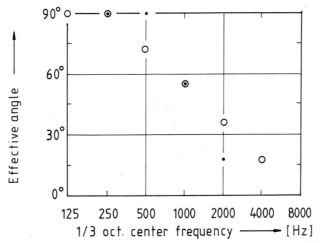

Fig. 5.6.9. Effective horizontal angles to a listener (for decreasing the IACC for the frequency band). Open circles, calculated; solid circles, measured.

factors. Binaural effects may also contribute to speech intelligibility and speech clarity. To examine the effects of the spatial factor on speech identification, speech intelligibility tests were conducted using a synthesized sound field and changing the delay time of the single reflection while the STI was kept constant (Nakajima and Ando, 1991). To obtain a range of speech intelligibility suitable for examining the effects of temporal and spatial factors, each monosyllable was joined by meaningless forward and backward maskers. The results shown in Fig. 5.6.10 indicate that the speech intelligibility increases with increase of the horizontal angle to a listener as well as with decreasing delay time. To explain the spatial effects, there are two simple models that need to be discussed.

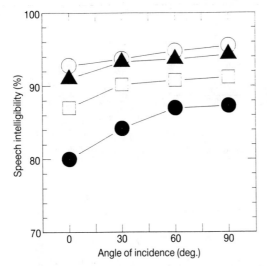

Fig. 5.6.10. Speech intelligibility for sound fields with a single reflection as a function of the horizontal angle to a listener and as a parameter of the delay time (6 subjects). Different symbols indicate different delay times of a single reflection Δt_n: $\bigcirc$, 15 ms; $\blacktriangle$, 30 ms; $\square$, 50 ms; $\bullet$, 100 ms. From Nakajima et al. (1991).

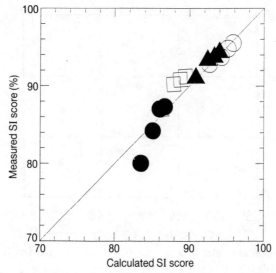

Fig. 5.6.11. Relation between measured SI scores and those calculated by using Eq. (5.6.5). Different symbols indicate different delay times of a single reflection Δt_1: $\bigcirc$, 15 ms; $\blacktriangle$, 30 ms; $\square$, 50 ms; $\bullet$, 100 ms. From Nakajima et al. (1991).

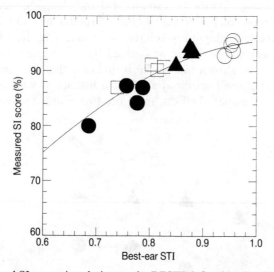

Fig. 5.6.12. Measured SI scores in relation to the BESTI defined by Eq. (5.6.6). Different symbols indicate different delay times of a single reflection Δt_1: $\bigcirc$, 15 ms; $\blacktriangle$, 30 ms; $\square$, 50 ms; $\bullet$, 100 ms. From Nakajima et al. (1991).

The first is the IACC model as described by

$$\text{SI} = f_1(t) + f_2(s) \tag{5.6.5}$$

where

$$f_1(t) = 111t + 15t^2 + 50t^3 + 18, \qquad t = \text{STI}$$

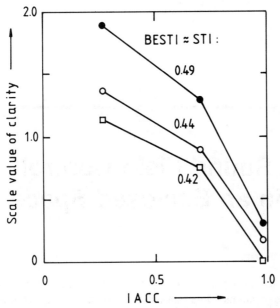

Fig. 5.6.13. Scale values of speech clarity as a function of the IACC and as a parameter of STI. (The STI is almost the same as the BESTI in this experimental condition with symmetrical loudspeaker arrangements for the reflections.) From Nakajima (1992).

and

$$f_2(s) = 10.5s - 2.5, \qquad s = 1.0 - \text{IACC}$$

As shown in Fig. 5.6.11, the SI scores calculated using Eq. (5.6.5) agree well with the measured scores.

The second model introduces BESTI, which is defined by (Miyata, et al. 1991)

$$\text{BESTI} = \{\text{STI(L)}, \text{STI(R)}\}\text{max} \qquad (5.6.6)$$

where STI(L) and STI(R) indicate the STI at the two ears. As shown in Fig. 5.6.12, the measured SI scores may also be expressed by those calculated by using Eq. (5.6.6).

Which of the two models is more closely related to speech intelligibility in a sound field? An experiment to evaluate clarity judgements associated with speech intelligibility was tried under conditions of varying IACC and fixed BESTI (Nakajima, 1992). Because these conditions are physically unreal for the usual sound field in a room, however, an experiment could not be performed under these conditions. As shown in Fig. 5.6.13 obtained for synthesized sound fields with a symmetrical loudspeaker system, however, the scale values of clarity obtained by paired-comparison tests (eight subjects) increase with decreasing IACC for fixed BESTI or STI; this change is significant ($p < 0.025$). The scale value of clarity is expressed by two independent factors, IACC and STI because there is no significant interference between these two factors.

6

Sound Field Control in an Enclosed Space

Probably the most useful aspect of digital signal processing is that it can be used to adaptively optimize systems. The acoustic echo-cancelling technique is a typical example of adaptive control. Chapters 6 and 7 present examples of sound field control technologies. In this chapter, we present basic technologies for practical applications. Problems with sound field control and optimization in a reverberant space are described for both wave theory and signal processing techniques.

6.1. Passive Control for Sound Power Transmission

As we demonstrated in Chapters 3 and 4, the sound power transmission into a reverberant space from a source varies with source location. Yanagawa (1988) pointed out that this phenomenon is seen in the vehicle interior sound field created by a car audio system. We will describe the sound source locations required for a flat power response, as for a car audio system, and also the sound power measurement of a sound source.

6.1.1. Sound Source Arrangement in an Enclosed Space

A. *Frequency Response Measurement*

Two loudspeakers are located either at both sides of the rear deck, or at both sides of the dashboard. The sound receiving points are at the driver's position, the front seat, and the back seat. A 1/3-octave band noise signal is fed into the two loudspeakers under an inphase condition. Figure 6.1.1 shows the frequency responses. When the sound sources are located on the rear deck, we can clearly see a dip at around 500 Hz at each receiving point. This type of dip in the mid-frequency range causes poor audio reproduction. As we explain in the next section, this dip results from the power response change due to the reflective surroundings.

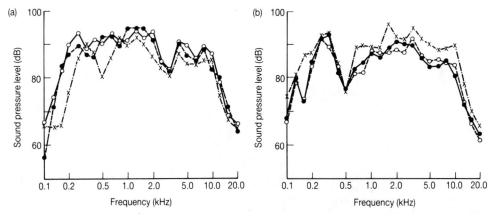

Fig. 6.1.1. Sound pressure responses at different observation positions: ○, driver's position; ●, front seat; ×, back seat. (a) Sound sources on the dashboard; (b) sound sources on the rear deck. From Yanagawa (1988).

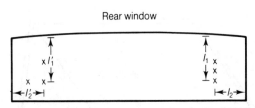

Fig. 6.1.2. Sound source positions on the rear deck for a second set of measurements. From Yanagawa (1988).

B. *Source Position Effects*

Figure 6.1.2 shows the sound source positions used for taking a second set of measurements. In this case, only one sound source is used and the sound receiving position is at the front seat. Figure 6.1.3 shows the measured results. We can still see a dip, but its frequency has changed owing to the different source location. Since the same type of dip is seen not only for two sound sources but also for one, we can conclude that this phenomenon is due to the power response changes caused by the reflective surroundings.

C. *Image Theory*

As mentioned in Section 3.2, power response can be estimated using image theory following Waterhouse (1963) for the case of simple surroundings. The geometrical configuration of the rear deck can be simplified as shown in Fig. 6.1.4. The angle ϕ is assumed to be 45° in order to apply the image theory, which can only be applied in this case when π/ϕ is an integer. Based on these assumptions, we only have to estimate

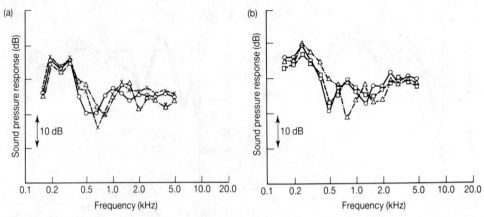

Fig. 6.1.3. Sound pressure responses (receiving point, front seat) for different sound source positions. (a) ○, $l_1 = 26.5$ cm, $l_2 = 19.5$ cm; ×, $l_1 = 21.5$, $l_2 = 19.5$ cm; △, $l_1 = 16.5$ cm, $l_2 = 19.5$ cm. (b) ○, $l'_1 = 26.5$ cm, $l'_2 = 19.5$ cm; △, $l'_1 = 16.5$ cm, $l'_2 = 19.5$ cm; □, $l'_1 = 26.5$ cm, $l'_2 = 9.5$ cm. From Yanagawa (1988).

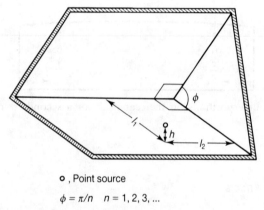

○ , Point source

$\phi = \pi/n \quad n = 1, 2, 3, \ldots$

Fig. 6.1.4. Model for the rear deck. From Yanagawa (1988).

the effects of 15 image sources. There are seven images in the plane encompassing the sound source and parallel to the side wall, and eight images are located at their image positions with respect to the side walls (see Fig. 6.1.5). As we described in Sections 3.2.2 and 3.2.3, the power output of a source with N image sources is given by

$$P_r = \sigma P_0, \qquad \sigma = 1 + \sum_{i=1}^{N} \frac{\sin kr_i}{kr_i} \qquad (6.1.1)$$

where r_i denotes the distance between the real source and the ith image source, and P_0 denotes the power output of the source in a free field.

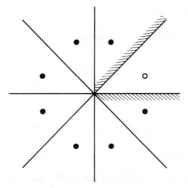

Fig. 6.1.5. Arrangement of image sources in the rear deck model. ○, Point source; ●, mirror images. From Yanagawa (1988).

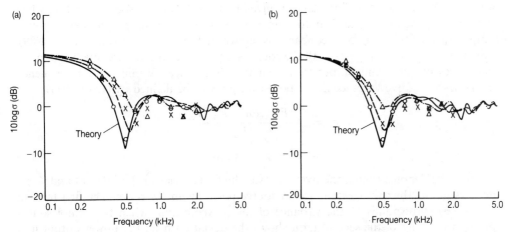

Fig. 6.1.6. Measurement of the power response in the model for different sound source positions. (a) ○, $l_1 = 26.5$ cm, $l_2 = 19.5$ cm; ×, $l_1 = 21.5$ cm, $l_2 = 19.5$ cm; Δ, $l_1 = 16.5$ cm, $l_2 = 19.5$ cm. (b) ○, $l_1 = 26.5$ cm, $l_2 = 19.5$ cm; ×, $l_1 = 26.5$ cm, $l_2 = 14.5$ cm; Δ, $l_1 = 26.5$ cm, $l_2 = 9.5$ cm. From Yanagawa (1988).

Figure 6.1.6 illustrates the calculated power responses for different source locations. We see a dip at around 500 Hz and the changes in dip frequency due to the source locations. This confirms that the dip in the frequency response is due to changes in the sound source power output. Since this dip is independent of the listener's position, we can electronically compensate for it using an equalizer. Power response control is also required for sound instruments located on a stage. The dip frequency becomes lower as the distance from the wall increases, while the depth of the dip becomes shallower. In general, therefore, it is recommended to locate a sound source or an instrument away from a flat reflective wall.

6.1.2. Sound Power Measurement

A. *Variance in Sound Source Power Response*

The sound power measurement methods are classified into several classes (ISO3740). For a small-sized source, a reverberation room is often used for the measurements: This is a very complicated task, since the power response of a source varies according to its location and frequency.

The power output P of a simple monopole source in a room is given by

$$P(x_s, \omega) = \tfrac{1}{2}\mathrm{Re}\{Z_{\mathrm{rad}}(x_s, \omega)\}Q^2 \quad \text{(W)} \tag{6.1.2a}$$

where Z_{rad} denotes the acoustic radiation impedance and

$$Z_{\mathrm{rad}}(x_s, \omega) = \frac{p(x_s, \omega)}{Q} = \frac{j\omega\rho c^2}{V}\sum_M \frac{\Psi_M^2(x_s)}{\omega^2 - \omega_M^2} \quad \text{(Ns/m}^5) \tag{6.1.2b}$$

where x_s shows the sound source position, p is the sound pressure at the source position (Pa), Q is the volume velocity of the source (m^3/s), and ρ is the volume density of the air (kg/m^3). The radiation impedance is given by the frequency response function when the receiving point is placed at the same position as the source. The sound power output varies depending on frequency, as shown in Fig. 6.1.7 (Tohyama et al., 1989), even if we take an average over all the sound source positions. We are interested in the differences of the sound output from that in a free space. Within a given frequency interval, the relative variance in the power output can be defined by (Lyon, 1969)

$$\sigma^2 = \frac{\langle\langle (P(x_s, \omega) - P_0)^2\rangle_{x_s}\rangle_\omega}{P_0^2}$$

$$= \langle\sigma_s^2(\omega)\rangle_\omega + \langle\sigma_0^2(\omega)\rangle_\omega \tag{6.1.3}$$

where $\langle\ \rangle_\omega$ denotes taking the average over the frequencies in the interval and $\langle\ \rangle_{x_s}$ denotes taking the average over sound source positions in the room. The first term in Eq. (6.1.3) corresponds to the variance of the power output when the sound source position is changed; the second term shows the deviation from the power output in a free field.

By introducing the assumption that the distribution of eigenvalue spacings is a Poisson distribution in the frequency band of interest (Section 3.5.4), the relative variances can be written as (Lyon, 1969; Davy, 1981; Tohyama et al., 1989)

$$\langle\sigma_0^2(\omega)\rangle_\omega = \frac{1}{2M}, \qquad \langle\sigma_s^2(\omega)\rangle_\omega = \frac{19}{16M}$$

$$\sigma^2 = \frac{1}{2M}\left(\frac{19}{8} + 1\right) \tag{6.1.4}$$

where M denotes the modal overlap. The relative variance in power output decreases as the modal overlap increases. To measure the power output of a pure tone source, we have to increase the modal overlap in the reverberation room (Suzuki and Tohyama, 1988).

The sound power measurement for a broadband noise source is more accurate than that for a pure tone source. Use of the sound pressure data observed on the boundary

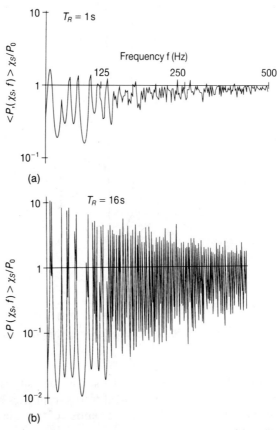

Fig. 6.1.7. Sound power output from a source averaged with the source position. P_0 is the power output of the source in the free field (W), the room volume is $200 \, m^3$, and the ratios of room dimensions are $1:2^{(1/3)}:4^{(1/3)}$. Oblique modes only are counted. From Tohyama et al. (1989).

of the reverberation room has been proposed in order to improve the accuracy for the broadband noise source (Tohyama and Suzuki, 1986).

B. *Sound Intensity Method*

Assuming that the dissipation of the sound energy in the medium is negligibly small, the radiated power from the source is obtained by measuring the sound intensity and integrating its normal component over a surface surrounding the source. The advantage of the sound intensity method is that it does not require a specially designed room such as an anechoic room or a reverberation room. The measurement surface is normally that of a semi-sphere or a rectangular prism. It is easily understood that the power obtained by integrating the normal component of the intensity over a surface is equal to the true power going through that surface (see Fig. 6.1.8). Let us assume that the sound energy flows normal to surface S_0 and the intensity I_0 is uniform over the surface. Then the power going through surface S_0 is given by $I_0 S_0$, which is equal to $I_x S_x$ since $I_x = I_0 \cos \phi$ and $S_x = S_0 / \cos \phi$.

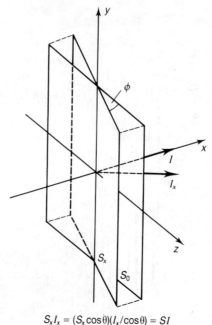

$$S_x I_x = (S_x \cos\theta)(I_x/\cos\theta) = SI$$

Fig. 6.1.8. Sound power going through a surface.

In order to be able to assume that the sound intensity is uniform over some area, the measurement surface must be divided into subsurfaces $S_n, n = 1, 2, 3, \ldots$, and then the sound intensity $I_n, n = 1, 2, 3, \ldots$ at the centres of the subsurfaces are measured. The radiated power is given by

$$P = \sum_{n=1}^{N} I_n S_n \quad \text{(W)} \tag{6.1.5}$$

This method is termed a discrete method. Results obtained by the discrete method can also be used to plot an intensity contour map. A disadvantage of the discrete method is that it is quite time consuming, since the intensity at each point is obtained after a number of averagings. In order to avoid this problem, the intensity probe is scanned over the measurement surface taking the areal and time averages continuously. The accuracy of this method has been confirmed experimentally.

An example of radiated power measurement and its application is shown in Figs. 6.1.9 and 6.1.10. Two identical loudspeakers are placed close to each other as shown in Fig. 6.1.9, which also shows measurement surfaces. Figure 6.1.10 shows the radiated power characteristics when (1) two loudspeakers are driven in-phase (w_1); (2) two loudspeakers are driven 180° out-of-phase (w_2); and (3) only one loudspeaker is driven (w_3). It is interesting that the radiated power is less when two loudspeakers are driven in-phase than when driven out-of-phase in some frequency ranges. From these radiated power measurement as well as the diaphragm velocity measurement, the self and mutual radiation resistances are experimentally determined (Suzuki et al., 1990).

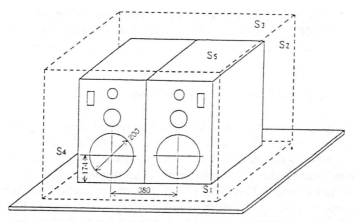

Fig. 6.1.9. An example of power measurement surface.

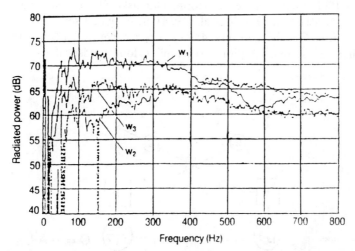

Fig. 6.1.10. Radiated power characteristics. w_1, two loudspeakers driven in-phase; w_2, two loudspeakers driven 180° out of phase; w_3, one loudspeaker only.

6.2. Sound Field Reproduction and Sound Image Control

Damaske (1971) and Shaw (1966, 1974) defined the head related transfer function (HRTF) in order to evaluate the relationship between sound image perception and the sound field conditions. The subjective diffuseness is also an important perception of spatial impression (Blauert, 1983). As discussed in Chapter 5, spatial impression is closely related to the interaural cross correlation function (ICF). Damaske and Ando (1972) defined IACC (interaural cross correlation) as the maximum absolute value of the ICF within a 1 ms time lag (see Eq. (5.1.8)).

We will investigate the interaural cross correlation coefficient (ICC), which is given by the value of ICF when the time lag is zero. The subjective diffuseness due to random reverberant sounds is also closely related to the ICC in a reverberant sound field.

6.2.1. Sound Image Control in a Reproduced Field

Stereo(phonic) reproducton was developed to reproduce an original sound field in a listening field. Itow (1957) and Makita (1962) analysed the sound field reproduced by two loudspeakers. Binaural stereophonic reproduction was developed to reproduce only the two signals that the listener hears in an original field in the listener's two ears (Shimada and Hayashi, 1992). As we will describe in Section 6.3, Schroeder and Atal (1963) proposed a system using cross-talk cancellers, as shown in Fig. 6.2.1. Cross-talk is the sound coming from the right (left) loudspeaker into the left (right) ear. The HRTF is a complex (phase and magnitude) function of the direction and frequency of a source. We define the HRTF in a free field as the complex difference between the sound pressure at the left (right) ear and the sound pressure at the listener's point when the listener is not present.

Using the HRTF as a complex function, Damaske (1971), Nakabayashi (1977), Sakamoto et al. (1981), and Gotoh et al. (1984), investigated sound image control in stereo reproduction. Gotoh et al. (1984) developed a control method that uses the concept of phantom source locations to simulate four loudspeakers, while actually

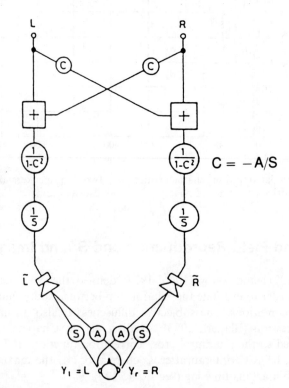

$$C = -A/S$$

Fig. 6.2.1. Binaural reproducing system using cross-talk cancellers. From Schroeder et al. (1974).

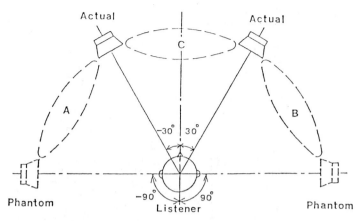

Fig. 6.2.2. Loudspeaker placement for stereo reproduction by a pair of actual loudspeakers. From Gotoh et al. (1984).

using only two loudspeakers, as shown in Fig. 6.2.2. The sound images in the regions A, B and C can be controlled using the level differences between two 'adjacent sources'. Good sound imaging is restricted to a 'sweet spot' in the reproduction area.

6.2.2. Expanding the Listening Area for Good Sound Image Localization

Improving the sound image localization for the stereo reproductions such as in an automobile environment and in teleconferencing is important, since the listeners are not on the centre line between the two loudspeakers. Sound image localization helps in identifying who is speaking in teleconferencing. In conventional two-channel stereo reproduction good localization occurs only in the small area that is equidistant from the two loudspeakers. The reproduced sound image tends to be localized at the nearer loudspeaker owing to the precedence effect (Haas, 1951, 1972). Aoki et al. (1990) proposed a new stereo system that expands the size of the listening area that has good localization. This method utilizes psychoacoustic phenomena and the directivity of a doublet source. Thus, the sound image localization is not very sharp, but it is also not highly sensitive to the listening position or listening room conditions.

A. Principle of the Method

The stereo reproduction method (Aoki et al. 1990) uses two sets of loudspeakers, each consisting of three individual loudspeakers, as shown in Fig. 6.2.3. The inner loudspeaker of each set (SP_r and SP_l) outputs the original two-channel stereo signals. Each loudspeaker signal is aimed at the opposite off-centre listening position. The outer loudspeaker of each set (SP_{ri} and SP_{li}) radiates the reversed phase of the original signal at an angle away from the signal of the inner loudspeaker. This pair of signals generates the radiation characteristics obtained by using a doublet source.

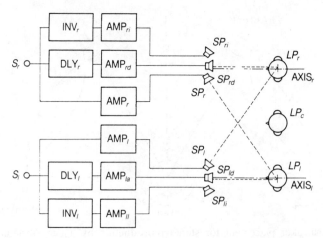

Fig. 6.2.3. Configuration of the proposed stereo reproduction method. INV, phase inverter; DLY, delay; AMP, amplifier; SP, loudspeaker; LP, listening position. Subscripts: l = left, r = right, c = centre, i = inverted, d = delayed. From Aoki et al. (1990).

The middle loudspeaker of each set (SP_{rd} and SP_{ld}) outputs a delayed original signal toward the off-centre listening position directly in front of the loudspeaker. The right signal is delayed so that it will reach a listener sitting directly in front of the right loudspeaker set simultaneously with the original signal from the left inner loudspeaker. The same timing relationship is used between the left delayed signal and the original signal from the right inner loudspeaker.

The effect of this stereo reproduction method is as follows. First, at the right listening position, LP_r, the original right signal from SP_r is accompanied by the phase-reversed right signal from SP_{ri}. In this case, the directivity of the doublet source reduces the sound pressure level. Even if the listener could perceive these sounds, the sound in his or her right ear is a phase-reversed signal of the original signal and the sound in his or her left ear is the original signal. The listener at LP_r will therefore get only a weak virtual image localization owing to the localization-in-head effect. With respect to the delayed signal, the listener hears the delayed right signal from SP_{rd} and the original left signal from SP_l, both of which contribute to the reproduction of the correct sound image. On the other hand, the delayed left signal from SP_{ld} does not contribute to the localization at LP_r owing to the precedence effect (Haas, 1951, 1972). Also, the phase-reversed left signal from SP_{li} is negligible since the acoustic axis of loudspeaker SP_{li} is pointed away from the listener.

At the centre listening position, LP_c, the listener first gets a virtual image of the original signals from the right and left inner loudspeakers, SP_r and Sp_i. Signals delayed several milliseconds from the right and left middle loudspeakers, SP_{rd} and SP_{ld}, are masked by the precedence effect. Phase-reversed signals from the right and left outer loudspeakers, SP_{ri} and SP_{li}, are negligible because their acoustic axes are pointed away from the centre position. By using the two sets of loudspeakers in this configuration, listeners positioned anywhere inside LP_r and LP_l can obtain the same localization as a listener positioned at LP_c.

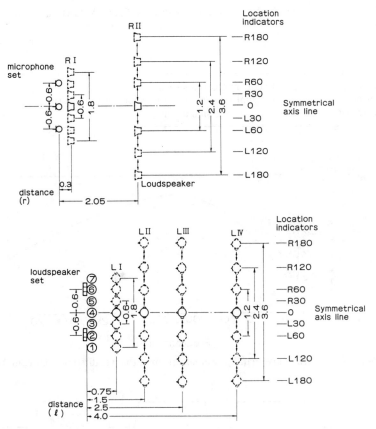

Fig. 6.2.4. Configuration for subjective test. (a) Recording conditions. (b) Listening conditions. From Aoki et al. (1990). Distances are measured in metres.

B. *Evaluation Experiments*

(1) *Preparation of Test Source for Subjective Test.* To test this proposed method, a subjective test was performed using a prerecorded source. A three microphone recording method was used. The microphone and loudspeaker positions for recording are shown in Fig. 6.2.4. The test signals were previously recorded in an anechoic room. A single loudspeaker represented a participant talking. This loudspeaker was set at the position (Fig. 6.2.4a) where the participant would sit (Fig. 6.2.4b).

The three microphones were set 0.6 m apart to each other and the output of the centre microphone was mixed with both channels, with its 0 dB relative level. Under the recording conditions shown in Table 6.2.1, at each of two distances from the microphone set, 0.3 m (RI) and 2.05 m (RII), seven different source positions, all in a line parallel to the microphones, were used. In Fig. 6.2.4a, R180 means that the loudspeaker is 180 cm from the symmetrical axis line of the microphone set. A weighted noise signal with a spectrum similar to that of a human voice was used as the talker's voice. The 400-Hz component was maximum, as shown in Fig. 6.2.5. The time window

Table 6.2.1. Parameters for the recording conditions. From Aoki et al. (1990)

	Recording conditions	
	R_I	R_{II}
Distance between left and right microphones (m)	1.2	1.2
Distance r between source signal loudspeaker and microphone sets (m)	0.3	2.05
Distance between positions of source signals (m)	0.3	0.6

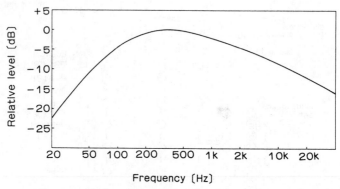

Fig. 6.2.5. One-third octave band power spectrum of the weighted noise. From Aoki et al. (1990).

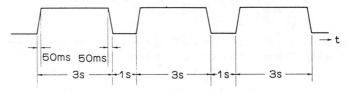

Fig. 6.2.6. Envelope of test signal. From Aoki et al. (1990).

of the test signal is shown in Fig. 6.2.6. The test signal consisted of three 3-s duration signals, including 50-ms rise and fall times and two 1-s pauses.

(2) *Listening Conditions for Subjective Test.* Subjective testing of the localization was conducted in an anechoic room. The configuration for the two loudspeaker sets and the various listener positions is shown in Fig. 6.2.4b. The reproduction conditions are shown in Table 6.2.2. The listener was positioned at either 0.75 m (LI), 1.5 m (LII), 2.5 m (LIII), or 4.0 m (LIV) from the loudspeaker sets, and at each of these four distances, at seven different locations to the right and left of the symmetrical axis line of the loudspeaker sets. Locations (R180, L180, etc.) are similarly defined in Fig. 6.2.4a. Subjects sat on a chair at a position selected randomly from among the listening positions in Fig. 6.2.4b. They then selected their sound image position from among the seven fixed locations (1 to 7) on a line running through the right and left loudspeaker sets. The subjects were not trained and they were allowed to move their heads to judge localization. Five subjects were selected for each test.

Table 6.2.2. Parameter for the listening conditions. From Aoki et al. (1990)

	Listening conditions			
	L_I	L_{II}	L_{III}	L_{IV}
Distance between left and right loudspeaker sets (m)	1.2	1.2	1.2	1.2
Distance *l* between loudspeaker sets and listening position (m)	0.75	1.5	2.5	4.0
Distance between listener positions (m)	0.3	0.6	0.6	0.6

C. *Results*

Figure 6.2.7 shows the subjective testing results for ordinary stereo reproduction and for this new method in an anechoic room. Figure 6.2.7a illustrates that with ordinary stereo reproduction, at the centre listening position, the relationship between real sources and virtual images is diagonal, that is, localization is good. At any other position, the relationships are not diagonal and source images are attributed to the location of either the right or the left loudspeaker set (R60 or L60) because of precedence effects. On the other hand, Fig. 6.2.7b shows that with the new method, the relationship between real sources and virtual images is diagonal within a listening location range up to 0.6 m from the symmetrical axis line. This means that within the listening area R60–L60, localization is good. Experiments similar to those described above were conducted and the same localization results were obtained in a reverberant room with a reverberation time of 0.7 s. On the basis of these results, configurations producing the best recording and listening areas for optimum localization using the new method are shown in Fig. 6.2.8.

6.2.3. Subjective Diffuseness and Interaural Cross Correlation in a Diffuse Field

As we described in Section 3.4, the sound pressure correlation coefficients for two points in a sound field are derived from the phase relationship between the two sound receiving points. Using these two points to represent the ears on a human head, it is possible to use these correlation coefficients to evaluate the sound field as heard by a human listener. If the sound field is a free field, the coefficients will correspond to the sound image localization resulting from the interaural time difference. In a diffuse field, they correspond to the perception that can be referred to as a subjective diffuseness resulting from the randomly changing time differences. In ordinary rooms, the correlation coefficient will correspond to both the localization and subjective diffuseness. Yanagawa and Anazawa (1990) have been investigating since 1969, as have Damaske (1967) and Damaske and Ando (1972), the relationship between the interaural cross correlation coefficient (ICC) and listener's subjective diffuseness.

A. *Cross Correlation Coefficients at Two Points in a Sound Field*

(1) *Correlation Coefficient in a Diffuse Field.* As we previously stated, the spatial correlation coefficients of sound pressure for two points can be written as

$$\rho_d(r) = \frac{\sin kr}{kr} \tag{6.2.1}$$

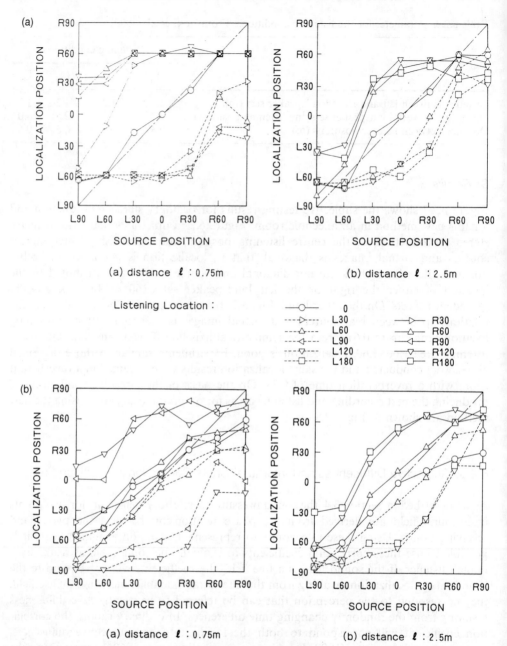

Fig. 6.2.7. Results of subjective tests when distance $r = 0.3$ m. (a) Ordinary stereo reproduction, (b) The proposed method. For example, R60 and L60 indicate areas that are 60 cm to the right and left of the symmetrical axis line. From Aoki et al. (1990).

where k is the wavenumber and r is the distance between two points. When undertaking actual measurements, it is convenient to make use of band-limited noise. If the frequency bandwidth is sufficiently narrow, it is possible to approximate Eq. (6.2.1) as $k = \omega_c/c$, where ω_c is the center angular frequency.

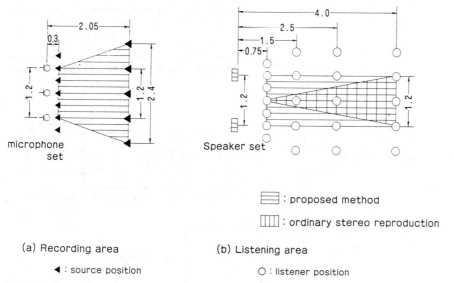

: proposed method

: ordinary stereo reproduction

(a) Recording area   (b) Listening area

◀ : source position   ○ : listener position

Fig. 6.2.8. Configuration of areas for good localization (all distances in metres). From Aoki et al. (1990).

(2) *Correlation Coefficient in a Mixed Field.* Concert halls and listening rooms, are considered to produce a mixed combination of free and diffuse sound fields. Under such circumstances, the correlation coefficient takes the form:

$$\rho_{df} = \left(\frac{K}{K+1}\right)\rho_d + \left(\frac{1}{K+1}\right)\rho_f \qquad (6.2.2)$$

In this equation, K is the ratio of the energies of the diffuse sound and the direct sound, ρ_d is the correlation coefficient for a diffuse field given by Eq. (6.2.1), and ρ_f is the coefficient for a free field,

$$\rho_f = \cos(kr\cos\theta) \qquad (6.2.3)$$

where θ represents the angle of incidence of the sound wave at the two receiving points.

B. *Experimental Methods*

The experimental arrangement is shown in Fig. 6.2.9. A 1/3-octave band noise was supplied to the reverberation room. The distance between the two recording microphones was varied, thus providing sample signals for various correlation coefficients. Utilizing the paired-comparison method for comparing values from two different samples, it was possible to isolate the subjective feeling of diffuseness and the difference in magnitude for this feeling from case to case. Judgements by test subjects were made under the dichotic listening condition. The test subjects determined, after listening to binaural sound over a headphone, which samples provided the greater amount of subjective diffuseness.

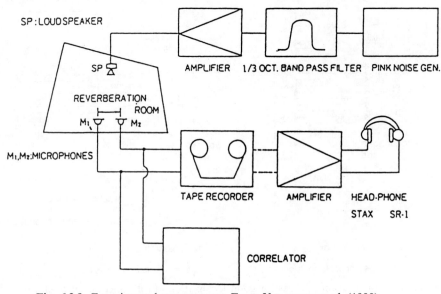

Fig. 6.2.9. Experimental arrangement. From Yanagawa et al. (1990).

The method of minimal changes was employed to derive the equiperceptual contours. The presented pairs consisted of the preceding reference sounds and the following sounds to be compared. The length of each sound was 2 s and the interval between the reference sound and the comparison sound was 1 s. The three possible answers in response to the question 'Did one of the sounds offer a greater feeling of diffuseness': were 'Yes', 'The same', and 'No'. A 1/3-octave band noise with centre frequencies of 125 Hz, 250 Hz, 500 Hz and 1 kHz was supplied.

Correlation measurements were made in a reverberation room with a volume of 102 m³ and a reverberation time of 3.5 s at 500 Hz. By varying the distance between recording microphones in the reverberation room, it was possible to produce diffuse sound samples with a ρ_d varying between 0 and 1.0 according to Eq. (6.2.1). When ρ_d ranged from -1 to 0, the antiphase was taken for the range of cases where ρ_d ranged from 0 to 1.0. Samples for other frequencies were not directly derived from the reverberation room, but rather by varying the dubbing speed of the 500-Hz recorded tape. By doing this, it was possible to limit the effect of diffuseness deterioration in the reverberation room over the lower frequency range.

In the final experiment, which considered the mixed sound field correlation coefficient ρ_{df} in Eq. (6.2.2), it was possible to obtain sample ρ_{df} values of 0.92 and 0.83 by mixing ρ_f of 1.0 and ρ_d of 0.15. Comparison could then be made with the ρ_d samples. In this experiment, 14 nontrained subjects and six professional recording engineers were employed. Their responses were consistent across both groups.

C. *Results*

The results shown in Fig. 6.2.10 represent judgements as to the existence or nonexistence of subjective diffuseness for various values of ρ_d in relation to a reference ρ_d of 0.3,

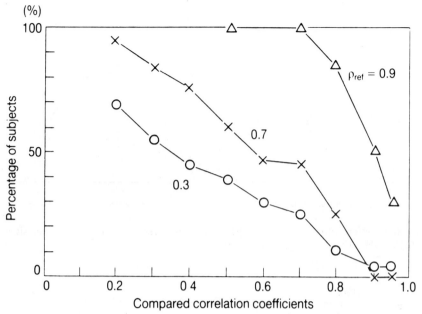

Fig. 6.2.10. Comparison of subjective diffuseness based on ICC. The percentage of subjects expressing that they felt a sense of diffuseness. ρ_{ref} = reference ICC. The centre frequency of 1/3-octave band noise is 1 kHz. From Yanagawa et al. (1990).

0.7 and 0.9. In the figure, the horizontal axis represents the ICC values, while the vertical axis shows the percentage of subject listeners indicating that they felt a sense of diffuseness. This figure makes it clear that there is a direct relationship between the ICC and subjective diffuseness. Going further, it can be seen that as ρ_d increases, subjective diffuseness tends to drop rapidly. This means that discrimination of the ICC is easier when the ICC value is large, as Gabriel and Colburn (1981) described.

Figure 6.2.11 shows the equiperceptual contour for the subjective diffuseness. In the case of Fig. 6.2.11a, a reference frequency of 1 kHz was employed and ρ_d was 0.3, 0.7 or 0.9. The reference frequency was 500 Hz in Fig. 6.2.11b, and ρ_d was 0.46 or 0.15. The broken curves in Fig. 6.2.11 show the results derived by Eq. (6.2.1) for the distance approximately equal to that between human ears, 32.8 cm (see Section 6.2.4A). When we compare the dotted lines to the experimental results of Fig. 6.2.11, it is interesting that the trends are similar.

The results presented thus far have all been for ρ_d in diffuse fields. Figure 6.2.12 shows results of combining free and diffuse fields. The reference frequency is 500 Hz, and the reference ICC(ρ_{df}) is 0.82 ($\rho_f = 1.0$ and $\rho_d = 0.15$) or 0.93 ($\rho_f = 1.0$ and $\rho_d = 0.15$). A comparison was then made with the subjective diffuseness for a diffuse field when the reference frequency was also 500 Hz. It can be seen that ρ_{df} also shows good correspondence with the subjective diffuseness. These results seem to indicate that both ρ_{df} and ρ_d can be considered a direct measure of subjective diffuseness. The perception of diffuseness is due to the binaural cue. If we listen to the noise signals independently in one ear and then the other, we are unable to detect any difference between them.

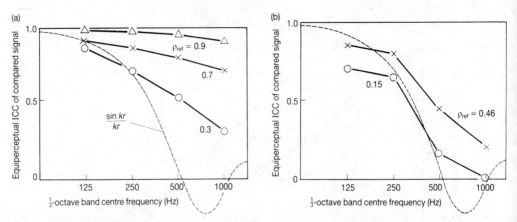

Fig. 6.2.11. Equiperceptual contour for the ICC. Centre frequency of the reference signal is (a) 1 kHz, (b) 0.5 kHz. From Yanagawa et al. (1990).

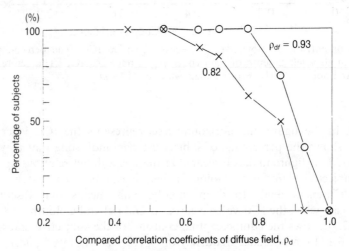

Fig. 6.2.12. Comparison of subjective diffuseness between mixed sound field and diffuse field based on ICC. ρ_{ref} = reference ICC of mixed field. The centre frequency of the reference noise is 500 kHz. From Yanagawa et al. (1990).

6.2.4. Reproduction of Subjective Diffuseness

Reproducing the spatial impressions of a concert hall is important for stereo recording and reproduction. This section examines ICC characteristics in two- and four-channel reproduced fields (Tohyama and Suzuki, 1986). These characteristics are compared with those in a diffuse field, since the reverberation field in a concert hall is regarded as a diffuse field. The relationship between the listener's preference and the reproduced sound field characteristics is treated in Chapter 5.

A. *ICCs of Sound Pressures in a Reverberation Field*

Suppose that white noise (or a band limited to a frequency band of interest) source is located in a concert hall. When a pair of sound pressure responses, $s_1(t)$ and $s_2(t)$, is observed at two receiving points, the point-to-point cross correlation coefficients $\rho_0(r)$ between those responses are given by

$$\rho_0(r) = \frac{\langle s_1(t) \cdot s_2(t) \rangle}{\sqrt{\langle s_1^2(t) \rangle} \sqrt{\langle s_2^2(t) \rangle}} \tag{6.2.4}$$

The distance between the two receiving points is r and $\langle \ \rangle$ denotes the ensemble average. Figure 6.2.13 shows the cross correlation coefficients measured in a large rectangular reverberation room. The distance between the two microphones is 15.2 cm, which equals the diametric distance between the ears of the KEMAR (Burkhard and Sachs, 1978). The measured data agree well with the theoretical curve by Eq. (6.2.1).

A listener's ICCs are different from the point-to-point cross correlation coefficients, because the sound waves are scattered or diffracted by the listener's head. Figure 6.2.14 shows the ICCs measured at the coupler microphone of the KEMAR. An eardrum simulator (B&K DB 100) with 0.5-inch type pressure microphones and small pinnas were used here. We found that the measured data follow the calculated curve of $\sin(kr)/kr$, if r is equal to 32.8 cm. This distance is the average effective distance, r_e, calculated from the cross correlation coefficients measured in each 1/3-octave frequency band from 100 to 1000 Hz. This effective distance is obtained by matching r_e to the measured $\rho(r)$ with a functional dependence given by

$$\rho_e(r_e) = \frac{\sin kr_e}{kr_e} \tag{6.2.5}$$

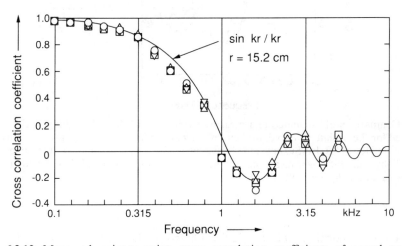

Fig. 6.2.13. Measured point-to-point, cross correlation coefficients of sound pressure in a rectangular reverberation room. Here, r is the distance between two microphones, 15.2 cm; k is the wavenumber (1/m); room dimensions are 11 m (length), 8.8 m (width), 6.6 m (height). From Tohyama and Suzuki (1989).

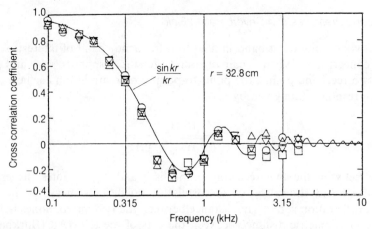

Fig. 6.2.14. Interaural cross correlation coefficients measured in a diffuse field using KEMAR. From Tohyama and Suzuki (1989).

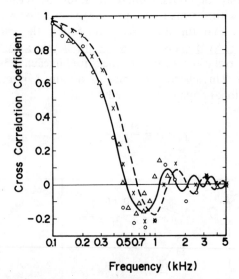

Fig. 6.2.15. Comparison of calculated and measured values of the ICCs in a diffuse field. Solid line, calculated for the prolate spheroid, broken line, calculated for the sphere (radius 7.5 cm). ○, measured for the KEMAR by Suzuki and Tohyama (1988); ×, measured for the sphere by Suzuki and Tohyama (1988); △, measured for the human head by Linvald and Benade (····). From Sugiyama and Tohyama (1989).

Consequently, we conclude that the 32.8 cm length is the effective acoustic distance between both the KEMAR's (or listener's) ears. This effective length is greater than that in a free field (Kuhn, 1977). Sugiyama and Tohyama (1989) analysed this equivalent interaural length using a prolate dummy head model. We can see a good correspondence between their theoretical estimation and the ICC experimental data shown in Fig. 6.2.15.

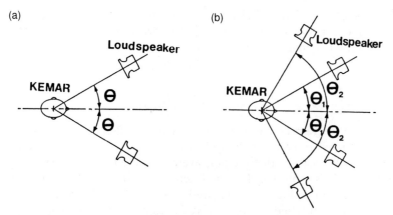

Fig. 6.2.16. Loudspeaker arrangements for stereo reproduction: (a) 2-channel, (b) 4-channel stereo. From Tohyama and Suzuki (1989).

B. *Reproduced ICCs in Two-Channel Stereo Reproduction*

As stated above, the ICC has the frequency characteristic of a sinc function in a diffuse field. The reproduced ICCs also depend on the loudspeaker arrangement. The loudspeaker arrangement in an anechoic room is shown in Fig. 6.2.16. Two 1/3-octave-band noise signals were fed to two loudspeakers in order to measure ICCs in the reproduced fields. Under conventional stereo recording conditions, the microphones for recording reverberation sounds are set far apart to prevent correlation between the recorded

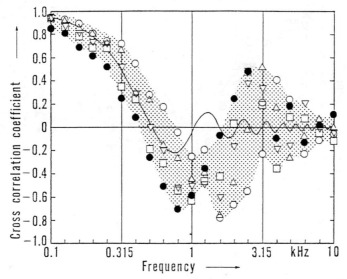

Fig. 6.2.17. ICCs in 2-channel stereo reproduced sound fields. (There is no correlation between the noise signals fed to the two loudspeakers.) $\bigcirc$, $\theta = 30°$; $\triangle$, $\theta = 35°$; ∇, $\theta = 40°$; $\square$, $\theta = 45°$; $\bullet$, $\theta = 5°$. Solid line is $\sin(kr)/kr$ where $r = 32.8$ cm. From Tohyama and Suzuki (1989).

reverberant sounds. The two loudspeakers were therefore driven with uncorrelated noise signals in order to simulate the recorded reverberant sounds.

The ICCs in two-channel stereo-reproduced fields are given by

$$\rho_{rep} = \frac{\overline{p_L(t) \cdot p_R(t)}}{\sqrt{\overline{p_L^2(t)}} \sqrt{\overline{p_R^2(t)}}} \tag{6.2.6}$$

where $p_L(t)$ is the sound pressure signal received at the left ear (Fig. 6.2.16), $p_R(t)$ is the sound pressure signal received at the right ear, and the overbar denotes a long time duration average. We assume that the time average equals the ensemble average. Figure 6.2.17 shows the ICCs reproduced at the listener's position in an anechoic room. The angle 2θ between the two loudspeakers (see Fig. 6.2.16) was changed from 60° to 100° in 10° steps. The solid line shows the ICCs calculated using $\sin(kr_e)/kr_e$, where $r_e = 32.8$ cm. From these results, we found that reproduced ICCs differed from those in the original reverberation field.

C. Reproduced ICCs in Four-Channel Stereo Reproduction

For four-channel stereo reproduction, the loudspeakers were located in a horizontal plane encompassing the listener, as shown in Fig. 6.2.16(b). All other conditions were the same as in the two-channel stereo reproduction experiments described above. Figure 6.2.18 shows the ICCs reproduced in an anechoic room by four-channel stereo reproduction. The solid line indicates the ICCs in the reverberation field.

The ICCs in four-channel stereo reproduction are closer to the ICCs in the original diffuse field. Variations in the ICCs with respect to the change in loudspeaker arrangement are relatively small for frequencies below 500 Hz. Above 500 Hz, the

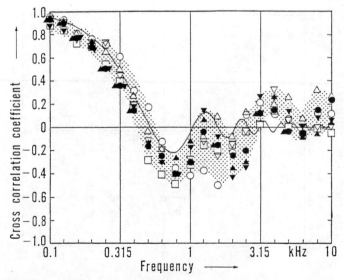

Fig. 6.2.18. ICCs in reproduced sound fields by 4-channel stereo reproduction. $\theta_1 = 30°$. ○, $\theta_2 = 45°$; △, $\theta_2 = 50°$; ▽, $\theta_2 = 55°$; □, $\theta_2 = 60°$; ●, $\theta_2 = 65°$; ▲, $\theta_2 = 70°$; ▼, $\theta_2 = 75°$. Solid line is $\sin(kr)/kr$ where $r = 32.8$ cm. From Tohyama and Suzuki (1989).

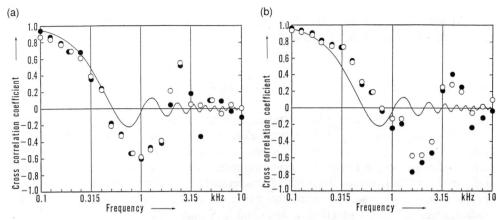

Fig. 6.2.19. ICCs in reproduced sound fields by symmetric-4-channel and by 2-channel stereo reproduction. (a) ●, $\theta = 45°$ (2-channel); ○, $\theta = 45°$ (4-channel). (b) ●, $\theta = 30°$ (2-channel); ○, $\theta = 30°$ (4-channel). Solid curves are $\sin(kr)/kr$, where $r = 32.8$ cm. From Tohyama and Suzuki (1989).

measured cross correlation coefficient ranges from approximately -0.45 to $+0.35$. In addition, the measured ρ does not show the fine structure of the correlation coefficient in a diffuse field at frequencies above approximately 500 Hz.

Figure 6.2.19 shows the ICCs in four-channel stereo-reproduced fields where the loudspeakers are arranged symmetrically in front of and behind the listener, as shown in Fig. 6.2.20. In this case, the ICCs are similar to those for two-channel stereo reproduction for frequencies greater than 500 Hz. It can be seen, then, that loudspeakers

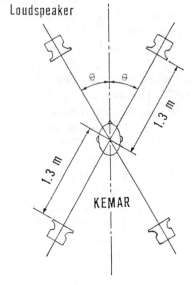

Fig. 6.2.20. Symmetric front and rear loudspeaker arrangements for 4-channel stereo reproduction. From Tohyama and Suzuki (1989).

should not be set symmetrically in front of and behind a listener if the ICCs of the reverberant field are to be reproduced.

6.2.5. Sound Field Simulator and Multichannel Reproduction

As we have discussed, sound field reproduction by the conventional stereo method is not completely satisfactory. Although the electroacoustic equipment used for recording and reproduction have been greatly improved, it is not easy to reproduce the spatial feeling of original sound fields in a listenting room. Subjective diffuseness is reproduced only in a very narrow area, even with four-channel stereo reproduction. Only minimal progress has been made in expanding the good sound image localization area for two-channel stereo reproduction as described in Section 6.2.2.

The concept of sound field synthesis has been proposed. This is a result of the rapid developments in using DSPs (digital signal processors) for audio applications. DSPs simulate the initial reflections of reverberation sounds in concert halls, and reproduce reverberation sounds in a listening room. The reverberation sound is a convolution between the source signal and the impulse response. DSPs made such real-time convolution processing possible. The added reverberation sounds in the reproduced field increase the subjective diffuseness perceived by the listener, thus giving the listener the feeling of being in a concert hall. The large negative ICC value produced in two-channel stereo reproduction (see Fig. 6.2.17), can be cancelled by the additional reverberation sounds, resulting in an expansion of the good listening area.

Figure 6.2.21 is a schematic diagram of a system by the Pioneer Electronic Corporation (1992). The workstation calculates the impulse responses using the image method of the room of interest. These simulated reflection sounds are convolved with the direct sound by a DASP (digital audio signal processor, DSP: PD0051) unit. The direct sounds are reproduced through two loudspeakers as in conventional two-channel stereo reproduction. The additional reverberation sounds, which are created by the convolution, are reproduced by six additional pairs of loudspeakers. These loud-speakers are located as shown in Fig. 6.2.21. Each pair of reverberation sound loudspeakers can reproduce a maximum of 480 reflections. The reverberation sounds, which are calculated by the workstation, are classified according to their direction and time delay, and assigned to one of the six pairs of reverberation channels.

The Yamaha Corporation (1992) has also developed a sound field simulation system using measured impulse response data in concert halls. This system can be applied not only to stereo reproduction for prerecorded sources, such as compact disks, but also to simulating the acoustic environment on a stage. This is a virtual environment technology. Figure 6.2.22 is a schematic illustration of this system. It is composed of four loudspeakers, one microphone, and a sound field processor. The microphone picks up the direct sound from the instrument and the received signal is fed to four independent FIR (finite impulse response) filters in the processor, which synthesizes four-channel reverberation signals by convolution. These FIR filters are constructed from the data measured using the closely-located four-point microphone method (Yamasaki and Itow, 1989), which will be introduced in Chapter 7. This measuring method was developed at Waseda University (Tokyo, Japan) and can analyse the direction of each reflection and synthesize the distribution of virtual sources. The reflections are classified into four groups corresponding to four loudspeakers. Yamaha's system can

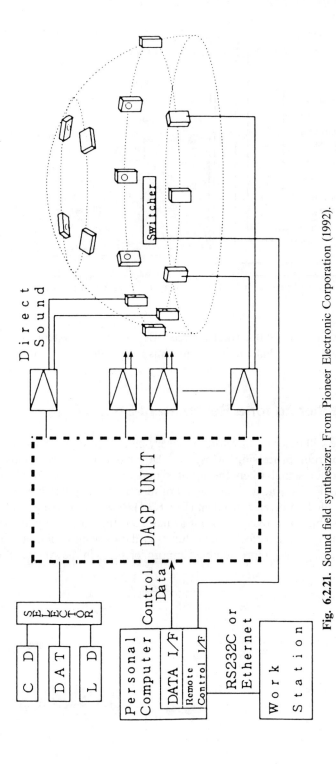

Fig. 6.2.21. Sound field synthesizer. From Pioneer Electronic Corporation (1992).

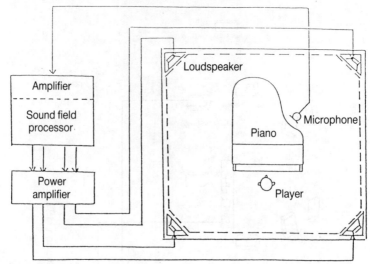

Fig. 6.2.22. Sound field processor for a piano player. From Yamaha Corporation (1992).

reproduce a maximum of 800 virtual sources. By using this system, musicians can practise in a virtual concert hall environment, thus improving their performance levels.

6.3. Inverse Filtering for a Reverberant-Sound Path

In this section, we will describe waveform recovery in a reverberant space using inverse filtering and cepstrum processing. Most conventional acoustic problems relate to estimating an output signal when the input signal and boundary conditions of the system are known. Recovering an input signal or identifying a system function from the output signal is called an inverse problem (Fig. 6.3.1). If we know the complete system function, it is possible to recover the input signal from the output signal. However, it is almost impossible to get perfect system function data owing to limitations such as the signal-to-noise ratio and unpredictable temporal changes in the system function. We

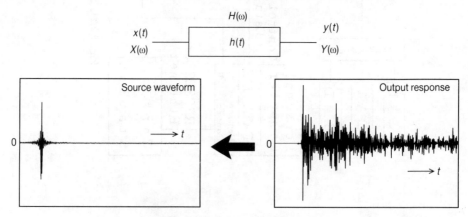

Fig. 6.3.1. Inverse problem.

have to estimate the input signal using incomplete system function data. Lyon (1987) is studying the application of inverse filtering to machine diagnostics.

6.3.1. Source Waveform Recovery in a Reverberant Condition

As we described in Chapter 3, the transfer function (TF) generally has nonminimum phase zeros, and the TF phase has a great variability from the theoretical trend. Inverse filtering is a challenge because of the non-minimum phase and complicated TF characteristics. In this section we introduce a method that takes the absolute value of the group delay and inversely filters only the minimum phase components of the TF in a reverberant space. The minimum-phase waveform can be recovered using the inverse filter for the minimum-phase part of the TF. Robust waveform recovery using a smoothed TF (Kim and Lyon, 1992) by using cepstrum windowing is also described.

A. *Ideal Inverse Filtering*

Recovering an input signal from the output response records is called inverse filtering. Suppose that $H(\omega)$ is the frequency response of a TF in a linear system. We can write the relationship between the output signal $Y(\omega)$ and an input signal $X(\omega)$ as

$$Y(\omega) = H(\omega)X(\omega) \qquad (6.3.1)$$

The input signal $X(\omega)$ can be recovered from $Y(\omega)$ as

$$X(\omega) = Y(\omega)/H(\omega) \qquad (6.3.2a)$$

$$= Y(\omega)\{1/H(\omega)\} \qquad (6.3.2b)$$

Here we call $1/H(\omega)$ an inverse filter of the linear system. It is impossible to create a causal and stable inverse filter of a TF that has nonminimum phase zeros. This is because the nonminimum phase zeros of the TF are inverted to the unstable poles of the inverse filter. It should be noted that Eq. (6.3.2a) still holds, even if the inverse filter $1/H(\omega)$ itself is not achievable. This is because all the nonminimum phase zeros of the numerator $Y(\omega)$ and the denominator $H(\omega)$ will, ideally, cancel each other out. We also note that even if the zeros in $H(\omega)$ and $Y(\omega)$ are minimum phase, they would still cancel. The inverse process above is possible under an ideal condition.

B. *Decomposition of Transfer Function into All-pass and Minimum Phase Components*

The TF of a nonminimum phase system can be written as the product of a minimum phase TF and an all-pass TF:

$$H(\omega) = H_{min} \times H_{\text{all-pass}} \qquad (6.3.3)$$

Suppose we have a TF expressed in terms of its pole-zero pattern, as in the upper part of Fig. 6.3.2. We can express this function as the TF shown in the middle of the figure, in which the non-minimum phase zeros have been reflected across the real frequency axis. This function therefore represents a minimum phase TF times a TF that has poles and zeros occurring only in pairs. The poles cancel out the new zeros introduced into the minimum phase TF. The TF represented by the pole-zero pattern in the lower part of the figure is an all-pass TF because the magnitude of the frequency response is unity. After creating the minimum phase part of the TF, we can obtain a stable inverse filter for the minimum phase part. The response waveform observed at a point in a system is changed from the source waveform by the phase characteristics of the TF. Therefore, the all-pass component of the TF is important for waveform recovery. The all-pass part is not reversible, however, because of its nonminimum phase property.

We now consider waveform recovery by inverse filtering only the minimum phase part of the TF. If a source waveform $X(\omega)$ itself does not have any nonminimum phase zeros, as shown by Fig. 6.3.3 and the TF does not change, we can recover the waveform by using only the inverse filter for the minimum phase part. As shown in Fig. 6.3.3, all the nonminimum phase zeros of $Y(\omega)$ are the zeros of the $H(\omega)$, and all the information about the source waveform is preserved in the minimum phase part of the TF. We can disregard the all-pass components of the TF for minimum-phase source waveform recovery.

C. *Minimum Phase Creation and Inverse Filtering*

The minimum phase decomposition described above is obtained based on the Hilbert transform as shown in Section 2.3. The minimum phase part of the TF can be also estimated by changing only the sign of the positive group delay components of the TF zeros. As shown in the pole-zero pattern (the middle part of Fig. 6.3.2), the nonminimum phase zeros of the original TF are reflected across the real frequency axis in the minimum phase part of the TF. The TF amplitude due to these symmetrical pairs of zeros is completely the same, while only the phase characteristics are affected by these zero locations.

The phase changes rapidly by π when the frequency passes near a zero. The sign of the phase jump near a zero depends on the location of the zero with respect to the frequency axis. When the zero is located below the frequency axis (called nonminimum phase zero), the sign of the π phase jump is negative (the group delay is positive). On the other hand, if the zero is located above the frequency axis (minimum phase), the sign of the phase jump is positive. Thus, the group delay of the TF phase due to these pairs of zeros has the opposite sign. Therefore, if the TF is truncated and the zeros are well separated in ω, taking the absolute value of the group delay due to TF zeros closely represents the group delay characteristics with the negative sign of the minimum phase part of the TF. When the zeros are densely distributed along the frequency line, the group delay functions overlap each other, and consequently the group delay is different from the case of isolated zeros.

(1) *Finite Impulse Response Record.* The signal processing procedure is explained in detail in Appendix 4. In brief, we take a truncated TF obtained from an impulse response record having a finite data length. All the TFs have only zeros in the discrete frequency plane and all poles are located at the origin of the discrete frequency plane.

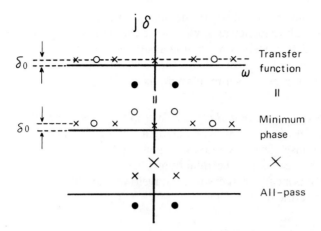

O: minimum phase zero

●: non-minimum phase zero

×: pole

Fig. 6.3.2. TF decomposition. From Tohyama et al. (1992a).

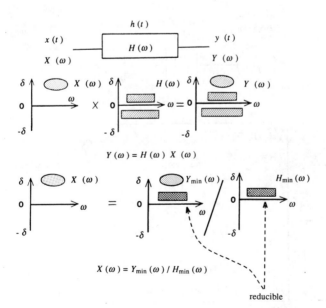

Fig. 6.3.3. Pattern of the pole and zero location. From Tohyama et al (1992a).

(2) *Minimum Phase Creation.* To move all the nonminimum phase zeros to symmetrical locations across the real frequency axis, we change the sign of the positive group delay part of the TF to negative. Here, we take the absolute value of the group delay. After taking the absolute value of the group delay, the newly obtained TF, which corresponds to the middle part of Fig. 6.3.2, is a minimum phase TF.

(3) *Inverse Filtering and Waveform Recovery.* We take a test response at the same receiving point used for the (reference) TF and again change the sign of the group delay part of the test response. The new minimum phase response is then inverted by the inverse filter already created for the minimum phase part of the reference TF. The minimum phase components of the source waveform are thereby recovered as the output signal by using the inverse filter.

(4) *Nonminimum Phase Component Recovery.* The nonminimum phase component can also be recovered by adding the phase difference between the all-pass parts of the test response and the reference TF.

D. *Recovered Waveforms in a Reverberant Space*

(1) *Reference TF Measurements.* Figure 6.3.4 shows the experimental arrangement. A series of measurements were made of the responses to a train of pulse-like source signals

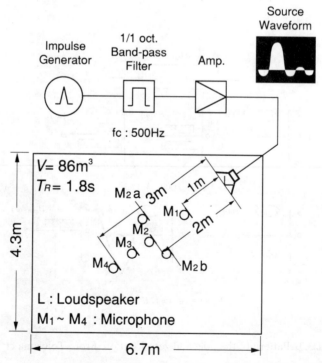

Fig. 6.3.4. Experimental arrangement for the source waveform recovery. From Tohyama et al. (1992b).

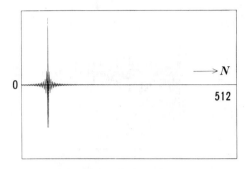

Fig. 6.3.5. Target source signal. From Tohyama et al. (1992a).

in an $86\,\mathrm{m}^3$ room with a reverberation time (T_R) of $1.8\,\mathrm{s}$. The interval between the pulse-like signals in the train has a sufficiently long duration compared with the reverberation time, so there is very little overlap. We assume that an ideal 1/1 octave band-pass filter was used to limit the frequency band of the pulse-like signal. The ideal source waveform, which we call the target signal, is shown in Fig. 6.3.5. We take one of the responses to the pulse-like signals as the reference impulse response $h_1(t)$ at M_1. Figure 6.3.6a shows only the initial 512 samples of the impulse response record. The record length for the impulse response completely covers the reverberation time of $1.8\,\mathrm{s}$. Figures 6.3.6b and c illustrate the amplitude, phase and group delay of the TF derived from the impulse response. The distance between the pole-line and real frequency line is about 4.0 $(\delta_0 = 6.9/T_R)$.

(2) *Inverse Filter for the Minimum Phase TF.* Figure 6.3.7 shows the amplitude and phase response of the minimum phase part obtained from the TF drawn in Fig. 6.3.6. This minimum phase part is extracted by taking the absolute value of the group delay shown in Fig. 6.3.6c. Figure 6.3.8 shows the impulse response of the inverse filter for the minimum phase response.

(3) *Samples of Recovered Waveforms.* Figure 6.3.9a shows the recovered waveform from the test response at M_1 when a pulse-like signal is radiated from the loudspeaker into the space. We used the inverse filter already created for the minimum-phase part of the TF at M_1. A greatly compressed waveform is extracted from the test response. The difference between the recovered waveform and the target source signal (shown in Fig. 6.3.5) is due to changes over time of both the TF and the source signal radiated from the loudspeaker. Figure 6.3.9b similarly shows a recovered waveform at M_3 from a test response at M_3 where we have created a new inverse filter using the reference response $h_3(t)$ measured at M_3. Waveform similar to that obtained at M_1 is recovered.

(4) *Nonminimum Phase Waveform Recovery.* A change of the minimum phase source signal into a nonminimum phase waveform can be detected. Figure 6.3.10 shows an example of a recovered source waveform that has an all-pass part. The difference from the original source waveform shown in Fig. 6.3.10 illustrates the change of the source waveform, if the TF is not changed over time.

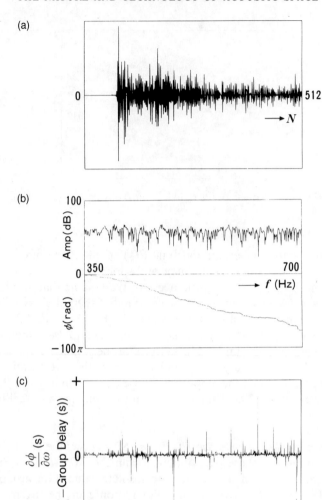

Fig. 6.3.6. (a) Reference impulse response, (b) TF magnitude and phase, and (c) group delay (first derivative of the phase). From Tohyama et al. (1992a).

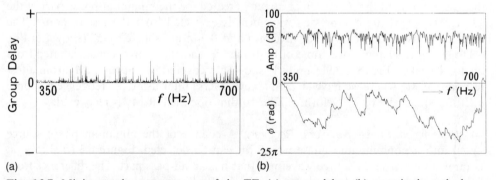

Fig. 6.3.7. Minimum phase component of the TF: (a) group delay; (b) magnitude and phase. From Tohyama et al. (1992a).

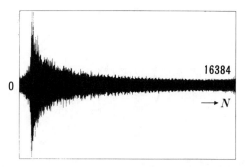

Fig. 6.3.8. Impulse response of the inverse filter. From Tohyama et al. (1992a).

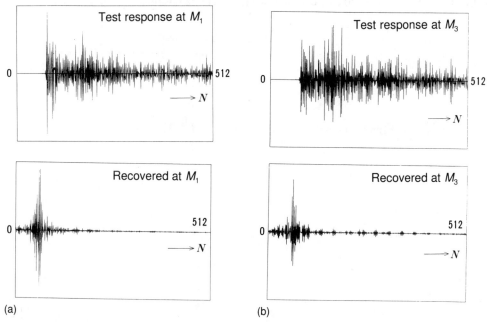

(a)   (b)

Fig. 6.3.9. Recovered source waveforms: (a) at M'_1, (b) at M_3. From Tohyama et al. (1992a).

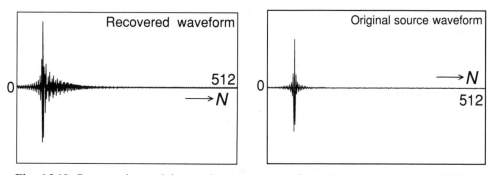

Fig. 6.3.10. Recovered nonminimum phase source waveform. From Tohyama et al. (1992b).

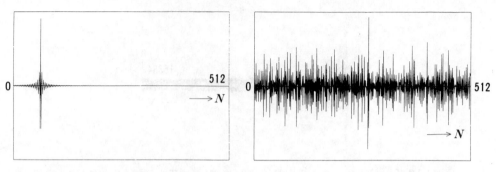

Fig. 6.3.11. Inverse filtering sensitivity: source waveform recovery at different positon. (a) Recovered reference waveform at reference position M_1 using the inverse filter created for M_1. (b) Recovered at M_3 using the inverse filter for the reference position M_1. From Tohyama et al. (1992a).

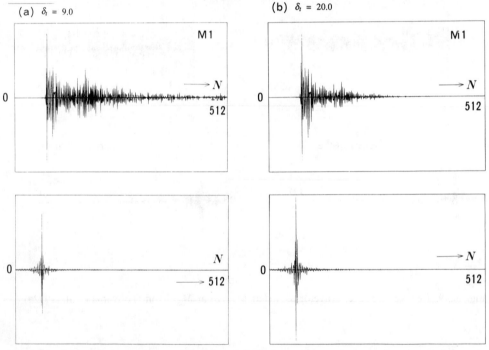

Fig. 6.3.12. Recovered source waveform using an exponential window. From Tohyama et al. (1992a).

E. *Exponential Windowing for Waveform Recovery*

Inverse filtering is sensitive to the location of the observation point. Since TF phase variation due to the observation locations is difficult to predict, we need a robust waveform recovery. If we take an inverse filter that was created using the reference response $h_1(t)$ at M_1 in order to recover the target signal from a test response at M_3, the waveform recovery does not work well. Figure 6.3.11a compares the same target signal to that of M_1; Fig. 6.3.11b illustrates a recovered signal from the test response at M_3 using the inverse filter for M_1. We can no longer clearly see the target signal.

Figure 6.3.12 is another example of recovered waveforms under different damping conditions. A similar source waveform is recovered using the newly created inverse filter under damping conditions. The changes in damping were caused by exponential windowing. Our waveform recovery results shown in Fig. 6.3.12 suggest that the source waveform can be recovered by exponential windowing without serious truncation effects, even if successive pulse train responses are observed in a reverberant space, as shown by Fig. 6.3.13.

F. *Smoothing a TF for Waveform Recovery*

(1) *Frequency Smoothing.* TF variations can be reduced by smoothing processes. Figure 6.3.14 shows examples of waveforms at M_{2a} and M_{2b} (Fig. 6.3.4) recovered by the inverse filter. This inverse filter was created by applying an exponential time window and a smoothing average in the frequency domain to the TF at M_2. Very similar waveforms can be recovered at M_{2a} and M_{2b}. The recovery process seems more robust to the unpredictable TF changes when the inverse filtering process smooths the test response.

(2) *Cepstrum Smoothing.* The complex cepstrum is defined as the inverse Fourier transform of the natural log of the TF. Figure 6.3.15 illustrates the relationship between the spectrum and the complex cepstrum (Kim and Lyon, 1992). Since the log magnitude is an even function of frequency and the phase is an odd function of frequency, their inverse transforms are real functions, so the complex cepstrum is a real function of time.

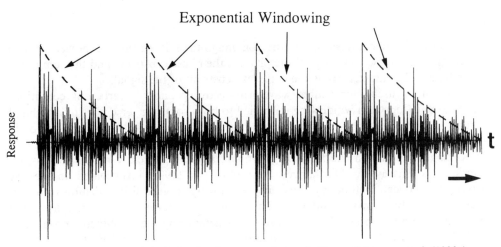

Exponential Windowing

Fig. 6.3.13. Exponential windowing for repeated signals. From Tohyama et al. (1992a).

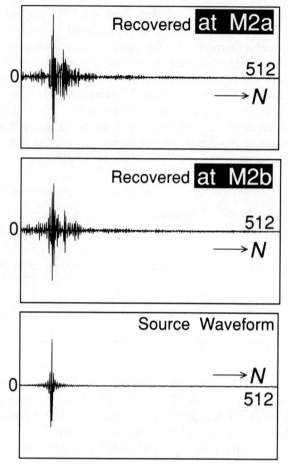

Fig. 6.3.14. Effect of exponential windowing and smoothing on the waveform recovery. Smoothing average, 60 Hz. Reduced effective T_R, 0.36 s. From Tohyama et al. (1992b).

The inverse Fourier transform of the log magnitude is called the power or real cepstrum. The inverse transform of the phase is the phase cepstrum, and the sum of the magnitude and phase cepstra is the complex cepstrum of the signal.

Cepstrum is useful for source waveform recovery under reverberant conditions (Lyon, 1987). Cepstrum windowing is applied to both robust inverse filtering and blind dereverberation. Because windowing in the time domain is equivalent to convolution in the frequency domain, the low-time windowing of the cepstrum is also equivalent to a smoothing operation in the frequency domain, and consequently makes the inverse filtering less sensitive to TF changes. Kim and Lyon (1992) investigated an effective waveform recovery method by cepstrum windowing. Figure 6.3.16 demonstrates an example of robust inverse filtering. The source waveform is recovered from the smoothed response observed at B by inverse filtering using the smoothed TF obtained at A. If we apply the inverse filter that was exactly made for the TF at point A, the recovery process no longer works well at point B, as shown in Fig. 6.3.17.

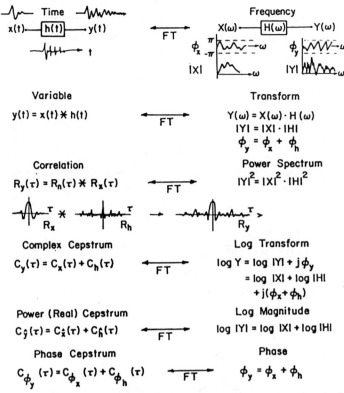

Fig. 6.3.15. The cepstrum has an even part (power cepstrum) and an odd part (phase cepstrum), each of which is additive in the time and frequency domains for source and path components. From Kim and Lyon (1992).

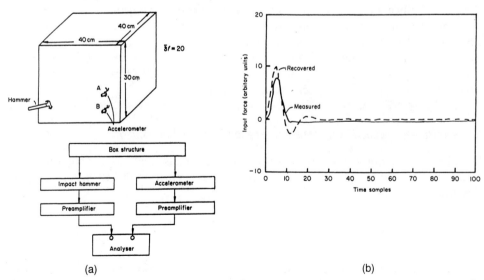

(a)   (b)

Fig. 6.3.16. Robust recovery using TF smoothed by cepstrum windowing. (a) Test structure for an impact waveform recovery. (b) Recovered waveform at B using the inverse filter for smoothed TF obtained at A. From Kim and Lyon (1992).

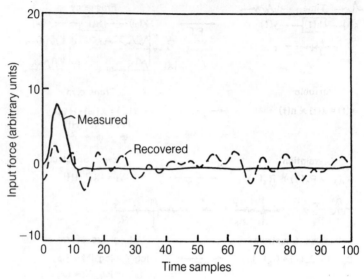

Fig. 6.3.17. Recovered waveform at B using the inverse filter for exact TF observed at A. From Kim and Lyon (1992).

6.3.2. Blind Dereverberation using Complex Cepstrum

Source waveform recovery is useful for machine diagnostics (Lyon, 1987) or speech processing (Wang and Itakura, 1992). We showed in the section above that the minimum phase components of pulse-like waveforms were recovered by inverse filtering of the minimum phase components of the TF. Waveform recovery by inverse filtering needs a reference TF. The TF phase trend is predictable (Tohyama et al., 1991) even under reverberant conditions, but inverse filtering is not stable because an individual TF phase will show large variations from the theoretical trend (Tohyama et al., 1991). In this subsection, we describe blind dereverberation without using a reference TF by applying complex cepstrum windowing (Yamasaki et al., 1982; Oppenheim et al., 1968). Complex cepstrum processing scheme is shown in Appendix 5.

A. *Blind Dereverberation using Minimum Phase Cepstrum Windowing*

We take a test response, and divide it into minimum phase and all-pass parts using the cepstrum. Following Eq. (6.3.3), the minimum phase cepstrum of the response $c_{Y\min}(\tau)$ is written as

$$c_{Y\min}(\tau) = c_{X\min}(\tau) + c_{H\min}(\tau) \qquad (6.3.4)$$

If the complex cepstrum of the source waveform $c_{X\min}$ has only low-time components compared to those of the TF $c_{H\min}$, the minimum phase source waveform can be recovered by low-time cepstrum windowing of the minimum phase cepstrum.

B. Recovered Waveforms

(1) *Cepstrum of a TF.* Figure 6.3.18 illustrates the all-pass and minimum-phase cepstra of TFs obtained at M_2, M_{2a} and M_{2b} in Fig. 6.3.4. The minimum phase cepstrum in the low-time region is not very sensitive to the observation location. Therefore we can expect to get nearly the same recovered waveforms at different locations, even if the fine structures of the TFs are changed.

(2) *Recovered Pulse-like Waveforms.* Figure 6.3.19 demonstrates waveforms recovered by windowing the low-time component (less than 10 ms) of the complex (total) cepstrum and minimum phase cepstrum of the response when a pulse-like signal is radiated from the loudspeaker into the space. Almost the same pulse-like wave-forms are extracted from the minimum phase cepstrum of the responses at different observation points. Figure 6.3.20 shows the effects of cepstrum window-length on the recovered waveforms.

C. Dereverberation of Speech

Blind dereverberation of reverberant speech has been tried by Yamasaki et al. (1982), but retaining the speech information remains a formidable problem. The power spectrum of a speech signal is essential for speech perception, but the nonminimum phase information is also needed to preserve the pitch of the speech waveform.

(1) *Zeros and Minimum-phase Components of Periodic and Non-periodic Waveforms.* Tim-ing information of the source signature is important. A periodic pulse train is an example of the timing information characterized by the minimum phase components, since all the zeros of the z-transform of the periodic pulse train are of minimum phase. Figure 6.3.21 is an example of the distribution of the zeros of a periodic pulse train. All the zeros are located on the real frequency axis (Hirobayashi et al., 1994). Figure 6.3.22 shows a sample of the minimum phase component of a piano tone (Hirobayashi, 1993). We can see no significant differences between the original and extracted minimum phase waveform. The pitch information of the piano tone can be conveyed by the minimum phase components.

A non-periodic train has non-minimum phase zeros in its z-transform, as we can see in Fig. 6.3.23. Pitch information is important timing information of a speech waveform. We need the non-minimum phase cepstrum in order to recover the speech waveform, since natural speech continuously change pitch. Figure 6.3.24 shows a minimum phase speech waveform recovered from the reverberant waveform. Comparing the recovered waveform with the original, we can detect spike-like noise due to the discontinuity produced by disregarding the all-pass part.

(2) *Time Window Length.* A speech waveform has two kinds of timing information: pitch and envelope. Let us take a time window length $2T_e$ that contains two periods of the speech envelope. The Fourier transform of this windowed signal can be written as

$$F_w(\omega) = F_e(\omega)(1 + e^{-j\omega T_e}) \qquad (6.3.5)$$

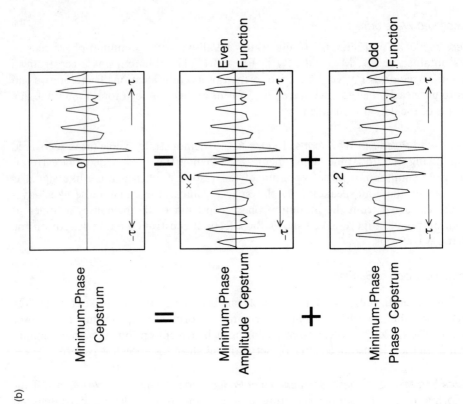

(b)

Minimum-Phase Cepstrum = Minimum-Phase Amplitude Cepstrum + Minimum-Phase Phase Cepstrum

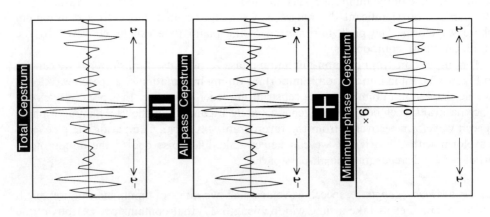

(a)

(c)

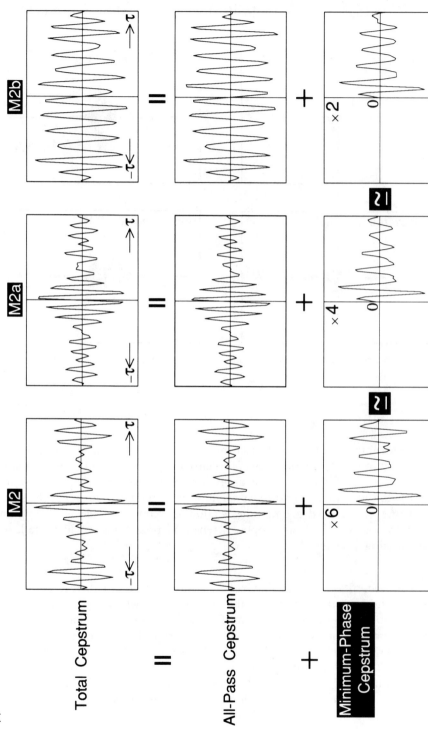

Fig. 6.3.18. Cepstrum decomposition and minimum phase cepstrum. (a) Cepstrum decomposition into all-pass and minimum phase cepstra. (b) Minimum phase cepstrum decomposition into magnitude and phase cepstra. (c) Cepstrum at M_2, M_{2a}, and M_{2b}.

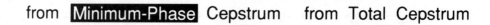

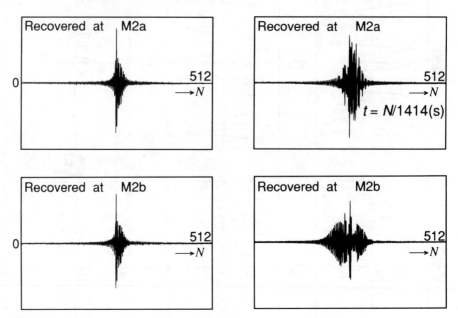

Fig. 6.3.19. Recovered waveforms by cepstrum low-time windowing. From Tohyama et al. (1992b).

where $F_e(\omega)$ denotes the Fourier transform of the speech signal in one period of envelope time duration. The periodicity of this envelope is represented by the minimum phase zeros of $1 + e^{-j\omega T_e}$. We cannot apply a very narrow cepstrum window to a long-time windowed speech waveform, as we have to keep the periodicity of the speech envelope. However, we need a long-time window to reduce the effects of long-time room reverberation. We therefore propose a repetitive processing in which the length of the time window is changed.

(3) *Recovered Speech Waveforms.* Figure 6.3.25 shows the original, reverberant and recovered speech waveforms recorded in the same room shown in Fig. 6.3.4. The effects of dereverberation are not pronounced, but Fig. 6.3.26 demonstrates that the reverberant decay components are slightly reduced by repeating the dereverberation process with varied time window length. Figure 6.3.27 shows the waveform D recovered by the single cepstrum windowing with $C_s = 0.125$ s as in recovery C applied to the same long speech time window $T_w = 1$ s as in Recovery A. Comparing with the recovered waveform C, we can see that the envelope of the recovered speech waveform D is distorted. We need to repeat the process with gradually reducing T_w and C_s.

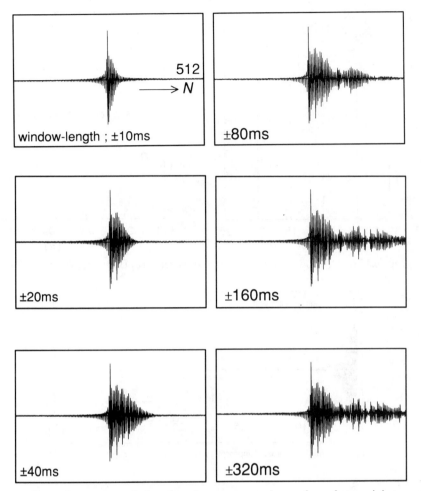

Fig. 6.3.20. Effect of cepstrum window length on recovered waveform from minimum phase cepstrum.

6.4. Active and Adaptive Control of Reverberant Sounds

Adaptive technology and active control for sound fields are reviewed in this section. The acoustic echo-cancelling technique is an example of adaptive control technologies. We will introduce a technology for an echo cancelling based on the energy decay property of reverberant sounds (Makino and Kaneda, 1992) and will describe systems that were proposed for sound field reproduction and control.

6.4.1. Acoustic Echo Canceller for Hands-free Telecommunication

A schematic diagram of a hands-free system is shown in Fig. 6.4.1. The acoustic echo canceller simulates the transfer function (TF) between a microphone and a loudspeaker

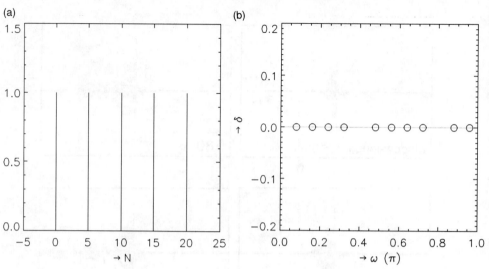

Fig. 6.3.21. A periodic pulse train (a) and its zeros (b) (Hirobayashi et al., 1994).

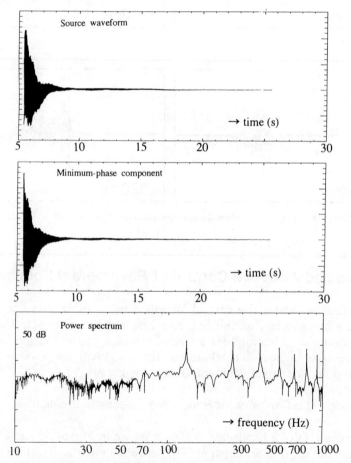

Fig. 6.3.22. Minimum-pulse extraction of a piano tone (Hiyobayashi, 1993).

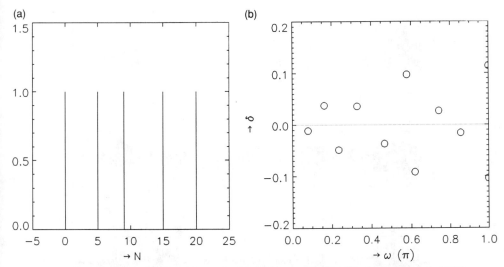

Fig. 6.3.23. A non-periodic pulse train (a) and its zeros (b) (Hirobayashi et al., 1994).

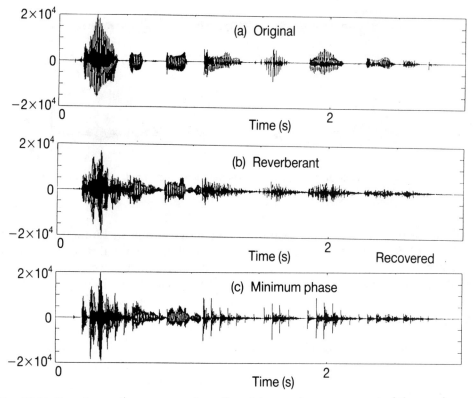

Fig. 6.3.24. Speech waveform recovery from the minimum phase component of the reverberant speech (Hirobayashi, et al., 1994).

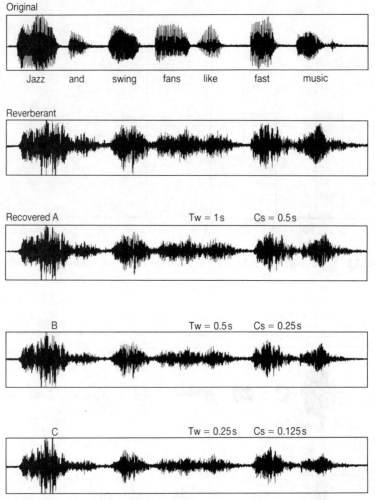

Fig. 6.3.25. Speech waveforms: original, reverberant and recovered. T_w = window length (s), C_s = cepstrum cutoff time (s). From Tohyama et al. (1993).

and cancels only the feedback signal by using the signal estimated from the echo-path filter (EPF). An adaptation process is needed to track the temporal changes of the TF. A conventional algorithm for this adaptation process is called the normalized least mean squares (normalized-LMS) algorithm (Nagumo and Noda, 1967; Haykin, 1991).

A. *Normalized-LMS Algorithm*

We can define the error signal to be corrected as

$$\varepsilon(n) = d(n) - \mathbf{h}^T(n + 1)\mathbf{x}(n) \tag{6.4.1}$$

where $d(n)$ is desired reponse and $\mathbf{x}(n)$ is an input signal vector, both measured at time n, and ^{T} denotes taking a transpose. The requirement is to find the impulse response vector (the tap-weight vector) $\mathbf{h}(n + 1)$ of the TFa or TFb in Fig. 6.4.1, measured at

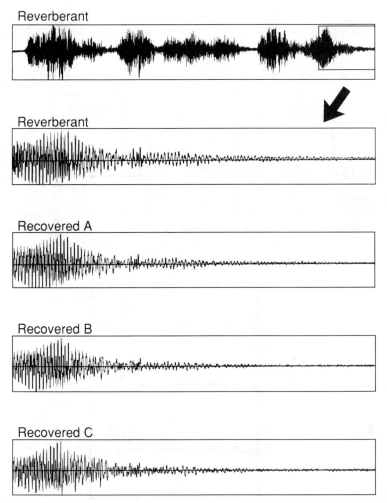

Fig. 6.3.26. Dereverberation effect on speech tail. From Tohyama et al. (1993).

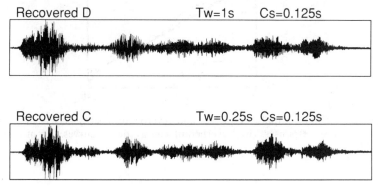

Fig. 6.3.27. Window length and cepstrum cutoff effect on recovered waveforms. From Tohyama et al. (1993).

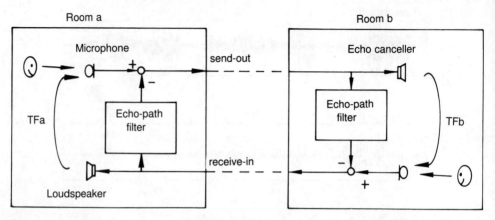

Fig. 6.4.1. Configuration of a hands-free telecommunication system. From Makino and Kaneda (1992).

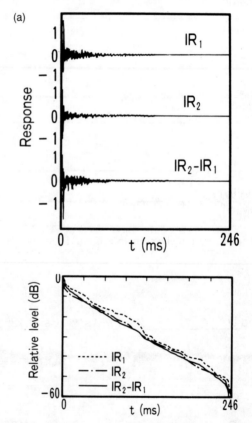

Fig. 6.4.2. Samples of room impulse responses and reverberant energy decay curves, and their differences. (a) Impulse responses. (b) Reverberant energy decay curves. From Makino and Kaneda (1992).

time n + 1, such that the change in the tap-weight vector given by

$$\delta\mathbf{h}(n + 1) = \mathbf{h}(n + 1) - \mathbf{h}(n) \tag{6.4.2}$$

is minimized, subject to the constraint

$$\varepsilon(n) = 0 \tag{6.4.3}$$

If we define the estimation error as

$$e(n) = d(n) - \mathbf{h}^{\mathrm{T}}(n)\mathbf{x}(n) \tag{6.4.4}$$

the constraint of Eq. (6.4.3) becomes

$$\mathbf{x}^{\mathrm{T}}(n)\delta\mathbf{h}(n + 1) = e(n) \tag{6.4.5}$$

Accordingly, our constrained minimization problem is equivalent to finding the minimum-norm solution for the change $\delta\mathbf{h}(n + 1)$ in the tap-weight vector at time $n + 1$, which satisfies the Eq. (6.4.5) (Haykin, 1991). We can get the solution as

$$\delta\mathbf{h}(n + 1) = \frac{\mu}{|\mathbf{x}(n)|^2} \mathbf{x}(n)e(n) \tag{6.4.6}$$

where μ is called a scaling factor and $|\mathbf{x}|^2$ is squared norm of $\mathbf{x}$.

B. *Fast Convergence using Exponentially-Weighted Step Size NLMS*

As described in Chapter 3, the TF are stochastic under high modal overlap reverberant conditions. In particular, since the arrangement of teleconferencing participants frequently changes during the conference, the sound field surrounding the echo canceller is time variant. The normalized-LMS generally does not use any statistical information about the system of interest. An effective EPF can be obtained if we can effectively use the statistics of the TF and can find an invariant property. Makino et al. (1991) has developed an adaptation procedure that utilizes TF statistics. As shown in Figure 6.4.2, Makino and Kaneda (1992, 1993) observed that the difference in the reverberation decay curves with the same damping coefficients decays exponentially as that of the averaged reverberation decay curve. From this finding, Makino and Kaneda developed a method that we call the exponentially-weighted step-size LMS algorithm. This new algorithm updates the tap-weight vector as

$$\delta\mathbf{h}(n + 1) = \frac{e(n)}{|\mathbf{x}(n)|^2} \begin{pmatrix} \mu_1 & \cdot & 0 \\ \cdot & \cdot & \cdot \\ 0 & \cdot & \mu_n \end{pmatrix} \mathbf{x}(n) \tag{6.4.7}$$

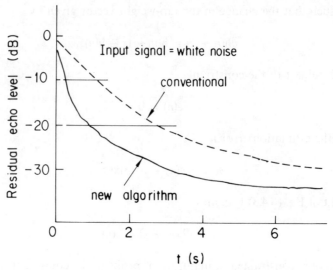

Fig. 6.4.3. Comparison of convergence. From Makino and Kaneda (1992).

where $\mu_i = \mu_0 \gamma^i$, $0 < \mu_0 < 2$, $0 < \gamma < 1$, and γ is the exponentially decay factor same as that of the reverberation decay curve. Figure 6.4.3 demonstrates that this new method reduces the adaptation time needed by the echo canceller compared to conventional ones. Many other trials and investigations of fast convergent performance have been reported (Fujii and Ohga, 1992).

6.4.2. Pole-zero Modelling of a Transfer Function

An FIR filter without (poles) has commonly been used for an echo-path filter. The number of FIR filter taps, however, must become very large as the reverberation time increases. More effective filter design is desirable for an audio echo canceller.

A. *Pole-Zero Model in Discrete-time Systems*

Transfer functions are characterized by poles and zeros. The TF can be given as a factored form in the z-plane,

$$H(z) = C \frac{\Pi_{k=1}^{M} (1 - c_k z^{-1})}{\Pi_{k=1}^{N} (1 - d_k z^{-1})} \qquad (6.4.8)$$

The total number of zeros is roughly of the same order as the number of poles.

B. *Experimental Samples*

Koizumi and Lyon (1989) measured the impulse responses of an acrylic cylindrical pipe and a small wooden box, and conducted pole-zero model identification (Fig. 6.4.4).

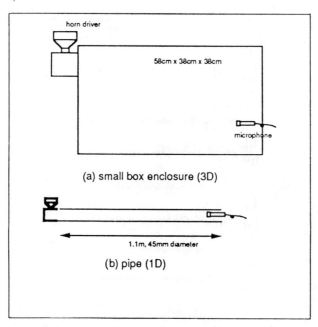

Fig. 6.4.4. Experimental models for pole-zero TF modelling. After Koizumi and Lyon (1989).

The cylindrical pipe was 45 mm in diameter and about 1 m long. The volume of the box was about $0.1 \, \text{m}^3$. A horn driver was used for excitation. In each experimental model, impulse responses were measured by generating maximum-length sequence pseudo-random noise and directly measuring the correlation function (Alrutz and Schroeder, 1983; Borish and Angell, 1983; Borish, 1985). Pole zero modelling was performed using the recursive identification algorithm (Ljung and Soederstroem, 1983).

(1) *Small Box Enclosure.* Figure 6.4.5 shows the measured impulse response (128 taps) for the small box with a sampling frequency of 1.8 kHz. The results for the 18-pole/18-zero model, the 16/16 and 12/12 models are shown. The 18/18 and 16/16 models agree well with the measured impulse response, but the 12/12 model fails to simulate the response. The pole-zero modelling is effective in reducing the FIR filter orders.

(2) *One-Dimensional Pipe.* The impulse response of the acoustical pipe (128 taps) is shown in Fig. 6.4.6. The results when the number of poles is reduced to 8 show that this number is insufficient for representing the response. On the other hand, the results when the number of zeros is reduced to 8 are similar to the upper two responses. This is because in the 1-dimensional case, when the loudspeaker and microphone are located at opposite ends, there are no zeros in the TF (Lyon, 1983). We can expect a large reduction in the number of filter taps in the very low-frequency region where only a few modes are excited (Haneda et al., 1994).

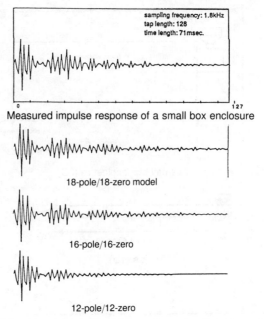

Fig. 6.4.5. Pole-zero model response for a small box enclosure. After Koizumi and Lyon (1989).

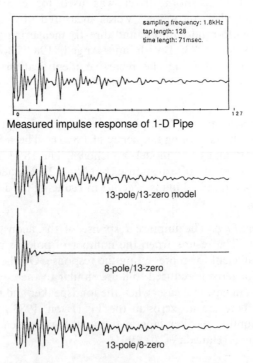

Fig. 6.4.6. Pole-zero model response for a 1-dimensional acoustic pipe. From Koizumi and Lyon (1989).

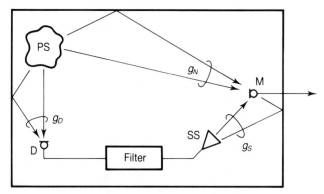

Fig. 6.4.7. Sound field control in a reverberant space. PS, primary sound (noise) source. SS, secondary source. From Miyoshi and Kaneda (1991).

6.4.3. Sound Field Control in a Reverberant Space

We will describe an application of number theory (Schroeder, 1985) to inverse filtering for sound field control (Miyoshi and Kaneda, 1988), and recent technologies for transaural recording and reproducing systems.

A. *Inverse Filtering for Sound Field Control—An Application of Diophantine Equations*

Active noise control is an example of sound field control. Figure 6.4.7 illustrates a situation for active noise control. There are two approaches to active noise control: one is to make a specific point M quiet and the other is to reduce the total radiated power from the noise source PS. These approaches are called quiet-zone control and active power minimization, respectively (Nelson and Elliott, 1991). The filter in Fig. 6.4.7 generates a noise cancelling signal to suppress the noise from PS at point M. The cancelling signal is synthesized from the primary source noise received by microphone D. This primitive configuration does not work well when the microphone D cannot be set close to the PS or when the SS cannot be set close to the noise control point M. This is because the detected noise signal or the controlling sound from the SS will be distorted by reverberation. Those distortion effects cannot be removed by the inverse filtering process, because a transfer function (TF) between two points has nonminimum phase zeros.

Figure 6.4.8 shows a schematic diagram for the multiple input/output inverse filtering theorem (MINT), which was developed by Miyoshi and Kaneda (1988). Two secondary sources, SS_1 and SS_2, and two filters, h_1 and h_2, are used to control the sound pressure response at point M. Inverse filtering requires that the filter input signal $x(n)$ and the output (observed at point M) signal $x(n - \tau)$ be identical except for the time delay τ. The filters must satisfy the equation

$$H_1(z)G_{s_1}(z) + H_2(z)G_{s_2}(z) = z^{-\tau} \qquad (6.4.9)$$

where z denotes the discrete complex frequency, G_{s_1} and G_{s_2} are the TFs, and τ is the sampled delay time. If these TFs have records with finite duration and do not have any

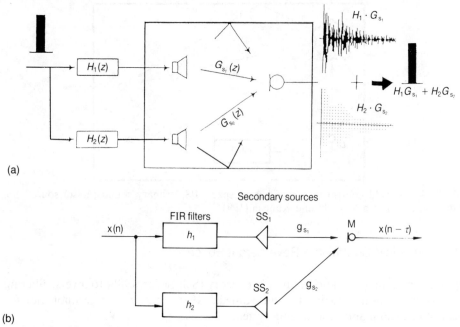

(a)

(b)

Fig. 6.4.8. Schematic diagram for sound field control based on MINT. After Miyoshi and Kaneda (1988).

common zero in the z-plane, then there exists a pair of stable FIR filters, H_1 and H_2. This MINT algorithm can be applied to an active noise control system for a reverberant space (Miyoshi and Kaneda, 1991). The sound pressures at N points can be controlled using $N + 1$ noise control units. Each control unit is composed of a microphone for detecting noise (the primary source signal), a filter, and a secondary source signal.

Equation (6.4.9) can be considered a Diophantine equation with respect to the polynomials (Schroeder, 1985). If $G_{s_1}(z)$ and $G_{s_2}(z)$ are coprime in the z-plane (there are no common zeros), then there exists a pair of stable FIR filters, h_1 and h_2. Figures 6.4.9a, b and c show the experimental results for creating a quiet zone around two noise-controlled points A and B in a reverberant space. The noise at point A is reduced by more than 25 dB when the noise signal extends from 50 to 400 Hz (Fig. 6.4.9b). Figure 6.4.9c illustrates the quiet zone created around the two points where the noise level reduction is more than 6 dB.

B. *Transaural System*

Sound recording using dummy heads and reproducing the original sounds by using two loudspeakers with cross-talk cancellers was originally proposed by Schroeder and Atal (1963) (see Fig. 6.2.1) (Schroeder et al., 1974). A binaural sound recording and reproducing system is known as the transaural system. In this section, we introduce transaural recording and reproduction using inverse filtering based on the MINT theorem (Miyoshi and Koizumi, 1991).

(a)

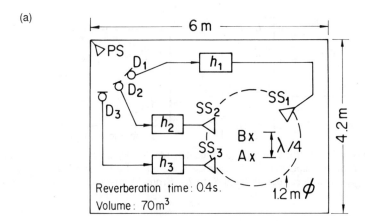

λ : wavelength of the center frequency
(λ = 152 cm. λ/4 = 38 cm)

Noise : 50-400 Hz band random noise

(b)

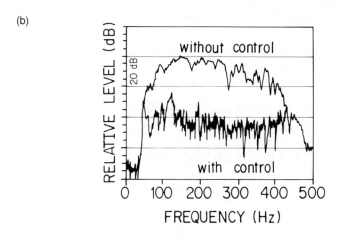

(c)

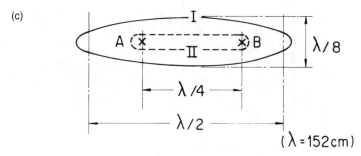

I : noise suppressed by over 6 dB

II : noise suppressed by over 14.5 dB

Fig. 6.4.9. Noise control experiment using MINT. (a) Experimental set-up. (b) Power spectrum of noise at noise-controlling point A. (c) Quiet zone created around two-controlling-points A and B. From Miyoshi and Kaneda (1991).

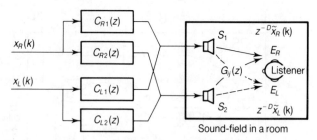

Fig. 6.4.10. Conventional transaural system using two loudspeakers. $x_j(k)$ $(j = R, L)$, input. $C_{ji}(i = 1, 2)$, filter. S_i, loudspeaker. E_j, listener's ear. $\tilde{x}_j$, reproduced sound. z^{-D}, modelling delay of D samples. $G_{ij}(z)$, transfer function from S_i to E_j. From Miyoshi and Koizumi (1991).

(1) *Conventional System.* Figure 6.4.10 shows a conventional transaural system. Acoustic signals $x_R(k)$ and $x_L(k)$ (k an integer index) are recorded at listener's ears in an original sound field. Signal $x_R(k)$ is emitted from loudspeakers S_1 and S_2 after it is filtered by $C_{R1}(z)$ and $C_{R2}(z)$, respectively. Signal $x_L(k)$ is likewise emitted after filtering by $C_{L1}(z)$ and $C_{L2}(z)$. The room transfer function from S_i ($i = 1, 2$) to E_j ($j = R, L$) is denoted by $G_{ij}(z)$.

Introducing the delay of D samples denoted by z^{-D}, filters $C_{j1}(z)$ and $C_{j2}(z)$ are calculated so as to satisfy the relationship:

$$z^{-D} = g_{1R}(z)C_{R1}(z) + g_{2R}(z)C_{R2}(z)$$
$$0 = g_{1L}(z)C_{R1}(z) + g_{2L}(z)C_{R2}(z)$$
$$0 = g_{1R}(z)C_{L1}(z) + g_{2R}(z)C_{L2}(z) \qquad (6.4.15)$$
$$z^{-D} = g_{1L}(z)C_{L1}(z) + g_{2L}(z)C_{L2}(z)$$

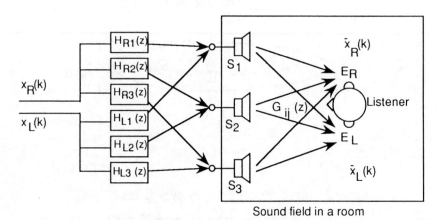

Fig. 6.4.11. Transaural system based on MINT. From Miyoshi and Koizumi (1991).

where $g_{ij}(z)$ is a polynomial obtained by removing from $G_{ij}(z)$ the group delay common to $G_{1j}(z)$, $G_{2j}(z)$. This set of equations gives the filters needed for achieving both the response equalization and the cross-talk cancellation introduced by Schroeder and Atal (1963). The performance of this inverse filtering method depends on the length of the delay D. Nelson, Hamada, and Elliott (1992) suggested that the required amount of delay may be about $1/2$ of the impulse response length.

(2) *Transaural System using More than Two Loudspeakers based on MINT.* The transaural system proposed by Miyoshi and Koizumi (1991) is shown in Fig. 6.4.11. This system uses three loudspeakers, S_i $(i = 1, 2, 3)$, and two sets of three filters, $H_{Ri}(z)$ and $H_{Li}(z)$, where filters $H_{j1}(z)$, $H_{j2}(z)$ and $H_{j3}(z)$ $(j = R, L)$ must satisfy

$$1 = g_{1R}(z)H_{R1}(z) + g_{2R}(z)H_{R2}(z) + g_{3R}(z)H_{R3}(z)$$
$$0 = g_{1L}(z)H_{R1}(z) + g_{2L}(z)H_{R2}(z) + g_{3L}(z)H_{R3}(z)$$
$$0 = g_{1R}(z)H_{L1}(z) + g_{2R}(z)H_{L2}(z) + g_{3R}(z)H_{L3}(z)$$
$$1 = g_{1L}(z)H_{L1}(z) + g_{2L}(z)H_{L2}(z) + g_{3L}(z)H_{L3}(z)$$

(6.4.16)

and $g_{ij}(z)$ is a polynomial obtained by removing from $G_{ij}(z)$ the group delay common to $G_{1j}(z)$, $G_{2j}(z)$ and $G_{3j}(z)$. MINT can uniquely determine $H_{j1}(z)$, $H_{j2}(z)$ and $H_{j3}(z)$ without introducing the delay D under the condition that there are no mutual zeros among the polynomials:

$$y_1(z) = g_{1R}(z)g_{2L}(z) - g_{1L}(z)g_{2R}(z)$$
$$y_2(z) = g_{2R}(z)g_{3L}(z) - g_{2L}(z)g_{3R}(z)$$
$$y_3(z) = g_{3R}(z)g_{1L}(z) - g_{3L}(z)g_{1R}(z)$$

(6.4.17)

Hence, the acoustic signals produced at listener's ears E_R and E_L are equivalent to signals $x_R(k)$ and $x_L(k)$.

(3) *Experimental Results.* The experimental apparatus is shown in Fig. 6.4.12. The reverberation times of the original field and the reproduced room are 0.2 s and 0.6 s, respectively. The room volume of both fields is $70 \, m^3$. In the original sound field, acoustic signals $x_R(k)$ and $x_L(k)$ are recorded at ears E_R and E_L of an artificial head when a loudspeaker is placed at P in the horizontal plane. In the reproduced field, loudspeakers S_1, S_2 and S_3 are placed on the circumference of a circle around the dummy head. Figure 6.4.13 shows the experimental results measured at E_R and E_L in the low frequency range. The measured error signals $r_R(k)$ and $r_L(k)$ were

$$r_R(k) = x_R(k) - x'_R(k), \qquad r_L(k) = x_L(k) - x'_L(k)$$

(6.4.18)

where $x'_L(k)$ and $x'_R(k)$ denote acoustic signals reproduced at the dummy head ears E_L and E_R, respectively. Curve (a) shows the power spectrum of $x_R(k)$. Curve (c) shows

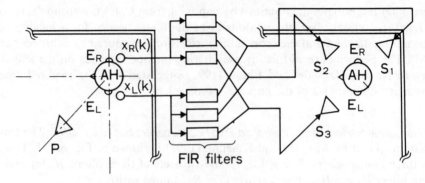

Original sound field   Reproduced sound field

Fig. 6.4.12. Experimental set-up for MINT transaural system. From Miyoshi and Koizumi (1991).

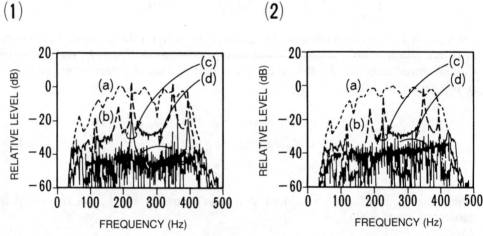

Fig. 6.4.13. Experimental results (1) at right ear E_R, (2) at left ear E_L. (a) Desired sound. (b) Errors by conventional system. (c) Errors by conventional system with a modelling delay of 400 ms. (d) Errors by proposed system. From Miyoshi and Koizumi (1991).

the error in the conventional system when a delay of 400 ms is introduced. Curve (d) is the reproduction error $r_R(k)$ in the MINT system. In MINT system, the error was about 30 dB less than the desired signal $x(k)$ at almost all frequencies. In the conventional system, however, some sharp peaks remain (see curves b and c). This is mainly because of the effect of the nonminimum phase zeros of the TFs.

6.4.4. Active Power Minimization in a Reverberant Space

Sound power output from a source can be reduced in a closed space by using secondary sources. These secondary sources must have a certain coherency with the primary source. This power-reduction method is called active power minimization. Nelson et al. (1987) investigated source locations for the best active minimization of enclosed sound fields. In this section, we will illustrate the power response from sound sources in a reverberation room and describe how the minimum power response (MPR) can be estimated using the approximate formula for power response (Tohyama and Suzuki, 1987) and the modal overlap in the sound field.

A. *Source Power Output in a Room*

The power output of a monopole source $P_0 \exp(j\omega t)$ located at a point p in a room is given by (Tohyama and Suzuki, 1987)

$$P_{ap} = \frac{P_0^2}{2} \operatorname{Re}\left[-j\, \frac{\omega\rho}{V} \sum_n \frac{\psi_n^2(p)}{(k^2 - k_n^2 - 2jk\delta/c)\Lambda_n} \right] \quad (\text{W}) \qquad (6.4.19)$$

where P_0 is the strength of the source (m³/s), ρ is the mass density of air (kg/m³), V is the room volume (m³), k is the wavenumber corresponding to the frequency of the source (1/m), k_n is the wavenumber corresponding to the nth eigenvalues of the room (1/m), δ is the imaginary part of the poles in the frequency interval of interest (1/s), ψ_n is the nth wavefunction (assumed real) of the room, Λ_n is the normalization constant for the nth wavefunction, and $\operatorname{Re}(x)$ denotes the real part of x. The sound power of a source can be derived from the space-averaged mean square pressure by (Tohyama et al. 1989)

$$P_{ap} = \langle p^2 \rangle \left(\frac{2V\delta}{\rho c^2} \right) \quad (\text{W}) \qquad (6.4.20)$$

where $\langle \ \rangle$ denotes a space average. We can define the power response of a source E_p as

$$E_p = \sum_n \left| \frac{\psi_n(p)}{k^2 - k_n^2 - 2jk\delta/c} \right|^2 \frac{1}{\Lambda_n} \qquad (6.4.21)$$

B. *Power Response with Secondary Sources*

The power response (TPR) from both a primary monopole source $P_0 \exp(j\omega t)$ located at point p and secondary monopole sources $Q_i \exp(j\omega t)$ located at points $q_m (m = 1, \ldots, I)$ is given by

$$E_p = \sum_n \left| \frac{P_0\phi_n(p) + Q_1\psi_n(q_1) + \cdots + Q_I\psi_n(q_I)}{k^2 - k_n^2 - 2jk\delta/c} \right|^2 \frac{1}{\Lambda_n} \qquad (6.4.22)$$

where I is the number of secondary sources. Suppose that N wave modes are excited in the room. TPR can be zero by using N secondary sources (Nelson et al., 1987). The number of modes, however, is generally unknown. The number of secondary sources is assumed to be smaller than the number of modes excited in the room. Therefore, TPR cannot become 0, but it does have a minimum value (MPR) (Nelson et al., 1987). This is called active power minimization.

C. MPR from a Primary and Secondary Source Pair

A monopole source $P_0 \exp(j\omega t)$ is located at point p, and a secondary monopole source $Q_i \exp(j\omega t)$ is located at point q in the room. TPR from both sources is

$$\text{TPR} = \sum_n \left| \frac{P_0\psi_n(p) + Q\psi_n(q)}{k^2 - k_n^2 - 2jk\delta/c} \right|^2 \frac{1}{\Lambda_n} \qquad (6.4.23)$$

We introduce an approximate formula for the power response to obtain MPR (Tohyama and Suzuki, 1987). The power response E_p under the low modal overlap can be expressed as the summation of the following two terms (Lyon, 1987):

$$E_p = D_p^2 + \varepsilon_p^2 \quad \text{where} \quad D_p^2 = \left| \frac{\psi_n(p)}{k^2 - k_n^2 - 2jk\delta/c} \right|^2 \frac{1}{\Lambda_n} \qquad (6.4.24)$$

The first term (due to the dominant resonance modes), D_p^2, expresses the resonance response contributed by the nearest resonance to the source frequency ($k \approx k_n$), and the second term, ε_p^2, expresses the remainder contributed by the other modes.

Introducing this approximation, TPR by Eq. (6.4.23) becomes

$$\text{TPR} \approx \left| \frac{P_0\psi_n(p) + Q\psi_n(q)}{k^2 - k_n^2 - 2jk\delta/c} \right|^2 \frac{1}{\Lambda_n} + P_0^2\varepsilon_p^2 + Q^2\varepsilon_q^2 \qquad (6.4.25)$$

if the secondary source is located far from the primary source. ε_q^2 expresses the remainder in the secondary source's power response E_q, due to off-resonance modes, and

$$E_q = D_q^2 + \varepsilon_q^2 \quad \text{where} \quad D_q^2 = \left| \frac{\psi_n(q)}{k^2 - k_n^2 - 2jk\delta/c} \right|^2 \frac{1}{\Lambda_n} \qquad (6.4.26)$$

The TPR takes on its minimum value (MPR) when the strength of the secondary source is

$$Q = \frac{-P_0\psi_n(p)\psi_n(q)}{\psi_n^2(q) + \varepsilon_q^2|k^2 - k_n^2 - 2jk\delta/c|^2\Lambda_n} \qquad (6.4.27)$$

The MPR then becomes

$$\text{MPR} \approx P_0^2 \left[E_p - D_p^2 \left(\frac{D_q^2}{E_q} \right) \right] \qquad (6.4.28)$$

The MPR depends on the location of the secondary source. The resonance response can be reduced to the remainder under the condition that $D_q^2/E_q \approx 1$. If the secondary source is located at the point where $D_q^2/E_q \approx 0$, it is impossible to reduce the power output of the primary source. If two secondary sources are used, the power response composed of two dominant resonance modes and a remainder can be minimized.

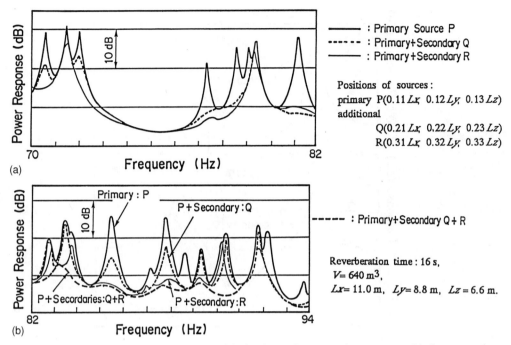

Fig. 6.4.14. Calculation of active power minimization using secondary sources. (a) One secondary source. (b) Two secondary sources. From Tohyama and Suzuki (1987).

Similarly, one can also develop the analysis to include cases in which n secondary sources are used.

D. *Examples of Calculated MPR*

For calculation, we assumed a $640\,\text{m}^3$ reverberation room. For simplicity, only oblique wave modes are taken into account. Figure 6.4.14a shows the calculated results for the MPR. The MPR was calculated numerically from Eq. (6.4.25) with the strength of the secondary source at each frequency given by Eq. (6.4.27). We found that some of the resonance response were reduced. Figure 6.4.14b illustrates the case in which two secondary sources are used. The power output is reduced at the frequency where the two dominant modes overlap.

E. *Experimental Results*

Figure 6.4.15a shows the power responses for pure tone sources in our reverberation room. The power response is reduced by a secondary source Q, located at a point far from the primary source P as shown in Fig. 6.4.16. The amplitude and phase of the secondary source are controlled in order to minimize the space-averaged mean square pressure level. The space-averaged level was estimated using mean square pressure data

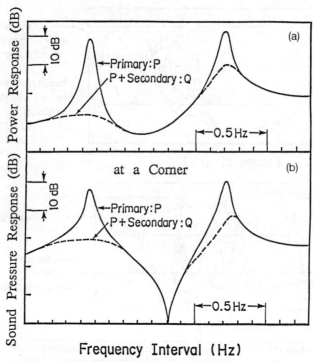

Fig. 6.4.15. (a) Power response and (b) sound pressure response at a corner from primary loudspeaker P and secondary loudspeaker Q. From Tohyama et al. (1991).

measured by six microphones located randomly. N sensor locations are necessary for active control in a room where N modes overlap. A corner, however, is a suitable sensor location when the modal overlap is very small or when uncorrelated modes are excited in the room. The sound pressure responses at a corner are illustrated in Figure 6.4.15b. The resonance peaks observed at the corner correspond to those of the power response, which is estimated from the space-averaged msp data. The power response can be minimized, if we control the amplitude and phase of the secondary source in order to minimize the sound pressure response data observed at the corner (Bartel et al., 1983; Tohyama and Suzuki, 1986b).

F. Estimation of MPR using Modal Overlap

The MPR has been evaluated using the dominant term and remainder. The space- and frequency-averaged remainder, can be evaluated using the modal overlap of the sound field. The expected ratio of the MPR to the power response of the primary source can be written as

$$\frac{\langle \text{MPR} \rangle}{P_0^2 \langle E_p \rangle} \equiv 1 - \frac{\langle D_p^2 \rangle}{\langle E_p \rangle} \frac{\langle D_q^2 \rangle}{\langle E_q \rangle} \approx 1 - \frac{1}{(1 + M)^2} \tag{6.4.29}$$

where

$$\frac{\langle D_p^2 \rangle}{\langle E_p \rangle} = \frac{\langle D_q^2 \rangle}{\langle E_q \rangle} \approx \frac{1}{1 + M} \tag{6.4.30}$$

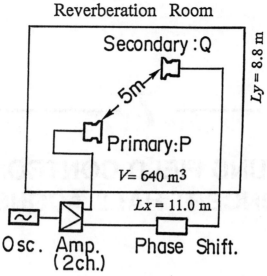

Fig. 6.4.16. Experimental arrangement. From Tohyama et al. (1991).

M is the modal overlap, $M = \pi \delta n_p(\omega)$, $n_p(\omega)$ is the modal density, $\delta = 6.9/T_R$, T_R is the reverberation time (s), and $\langle \ \rangle$ denotes a space average. M is about 0.25 for the conditions under which the numerical calculations of Fig. 6.4.14 were performed. In such a case, the expected level ratio is about 4.4 dB.

As shown in Figure 6.4.15, the resonance peaks are reduced using a secondary source. Equation (6.4.30) predicts an 8.5 dB reduction for our experimental conditions in Fig. 6.4.16. Nelson and Elliott (1991), proposed a more effective formula under the very low modal overlap.

7

SOUND FIELD CONTROL
FOR CONCERT HALL ACOUSTICS

Concert halls exemplify the effects of sound field control theories and technologies. This chapter presents technologies in measurement, evaluation, theory and design of concert halls. Speech intelligibility and sound detection are also described.

7.1. Sound Reinforcement for Speech Intelligibility

Speech intelligibility in a reverberant space like a concert hall is a primary objective for sound field control. We describe an estimation method for speech intelligibility, and technologies for loudspeaker (and/or microphone) arrays for improving speech intelligibility.

7.1.1. Estimation of Speech Intelligibility

A. *RASTI and MTF*

The modulation transfer function (MTF) and the rapid speech transmission index (RASTI) are physical properties for estimating SI scores (Houtgast et al., 1980; Houtgast and Steeneken, 1985). Here $h^2(t)$, the square of the impulse response, has two components:

$$h^2(t) = h_d^2(t) + h_r^2(t) \tag{7.1.1}$$

The direct sound term, $h_d^2(t)$, takes the form of a Dirac delta function, and reverberant sound term, $h_r^2(t)$, is an exponential decay function with a time constant, $T_R/13.8$. We assume that the onset of the exponential reverberation coincides with the arrival of the direct sound. We can express $h_d^2(t)$ and $h_r^2(t)$ as follows.

$$h_d^2(t) = \frac{q}{r^2} \, \delta(t) \tag{7.1.2}$$

and

$$h_r^2(t) = \begin{cases} \dfrac{1}{r_c^2} \dfrac{13.8}{T_R} \exp\left(\dfrac{-13.8t}{T_R}\right) & \text{for } t > 0 \\ 0 & \text{for } t \leqslant 0 \end{cases} \tag{7.1.3}$$

where $\delta(t)$: delta function
$\quad\quad r$: talker-to-listener distance (m)
$\quad\quad q$: directivity index of the sound source
$\quad\quad r_c$: $\sqrt{0.0032V/T_R}$ = the critical distance of the sound field (m)
$\quad\quad V$: the room volume (m^3)
$\quad\quad T_R$: the room reverberation time (s)

The modulation transfer function, $m(F)$, is given by Eq. 3.7.4. From Eqs (7.1.1) to (7.1.3), the modulation transfer funtion can be expressed as

$$m(F) = \frac{|\sqrt{A^2 + B^2}|}{C} \tag{7.1.4}$$

with

$$A = \frac{q}{r^2} + \frac{1}{r_c^2}\left[1 + \left(\frac{2\pi F T_R}{13.8}\right)^2\right]^{-1}$$

$$B = \frac{2\pi F T_R}{13.8}\frac{1}{r_c^2}\left[1 + \left(\frac{2\pi F T_R}{13.8}\right)^2\right]^{-1} \tag{7.1.5}$$

$$C = \frac{q}{r^2} + \frac{1}{r_c^2}$$

The RASTI can be calculated using the speech transmission indices calculated for the 500 Hz and 2 kHz octave bands, denoted STI$_{500\text{Hz}}$ and STI$_{2\text{kHz}}$ respectively.

$$\text{RASTI} = \frac{4\text{STI}_{500\text{Hz}} + 5\text{STI}_{2\text{kHz}}}{9} \tag{7.1.6}$$

where STI$_{500\text{Hz}}$ and STI$_{2\text{kHz}}$ are defined as

$$\text{STI}_{500\text{Hz}} = \frac{1}{4}\sum_i \frac{S/N_{\text{app},F_i} + 15}{30} \tag{7.1.7}$$

with modulation frequency F_i = 1, 2, 4 and 8 Hz, and

$$\text{STI}_{2\text{kHz}} = \frac{1}{5}\sum_i \frac{S/N_{\text{app},F_i} + 15}{30} \tag{7.1.8}$$

with modulation frequency F_i = 0.7, 1.4, 2.8, 5.6 and 11.2 Hz. The apparent signal-to-noise ratio, $S/N_{\text{app},F_i}$, in both Eq. (7.1.7) and Eq. (7.1.8) is given as

$$\frac{S}{N_{\text{app},F_i}} = 10\log\left(\frac{m(F)}{1 - m(F)}\right) \tag{7.1.9}$$

B. *Definition of Weighted MTF*

Miyata and Houtgast (1991) proposed a method that uses an exponentially decaying time window function for determining the MTF in a reverberant space. Intelligibility tests using a variety of computer-generated squared-impulse responses confirm that their method is better than the ordinary MTF (RASTI) method. The weighted MTF is expressed by

$$m_W(F) = \frac{\left| \int_0^\infty W(t)h^2(t)\, \exp(-2\pi jFt)\, dt \right|}{\int_0^\infty W(t)h^2(t)\, dt} \tag{7.1.10}$$

where

$$W(t) = \exp(-t/\tau) \tag{7.1.11}$$

and τ is a window parameter, for example, 0.2 ms. A weighted RASTI can be derived from the weighted MTF, by the same way as the conventional RASTI.

C. *Speech Intelligibility Tests*

(1) *Reverberation and Carrier Phrase Conditions.* Five echograms (squared impulse response) were prepared: four computer-generated echograms and one measured in a

Table 7.1.1. Models of echograms. After Miyata and Houtgast (1991).

$h^2(t)*$	Pattern	Parameter
Exponential	$e^{-(13.8t/T_R)}$	Reverberation time, T_R (s) 0.5, 1.5, 3, 4, 5, 6
Rectangular	b	Pulse width, b (s) 0.09, 0.22, 0.40, 0.46, 0.49, 0.65
Triangular	b	Pulse width, b (s) 0.12, 0.30, 0.55, 0.68, 0.80, 0.95
(Direct + exponential)	direct $e^{-(13.8t/T_R)}$	Distance between source and listening point, r (m) 1, 2, 3, 4, 6, 8 ($T_R = 3$ s)
Real field (reverberation room)	——————	Distance between source and listening point, r (m) 1, 2, 4, 6 ($T_R = 3$ s)

*$h(t) = \sqrt{h^2(t)} \times (M\text{-sequence}) = $ impulse response.

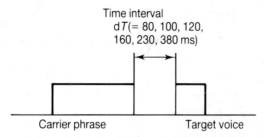

Fig. 7.1.1. Time sequence of the test signal used for speech intelligibility tests. From Miyata and Houtgast (1991).

reverberant field. Table 7.1.1 shows these echograms. The impulse response of each echogram was obtained by extracting the square root of the echogram and multiplying it by the maximum length sequence (Schroeder, 1985) noise in order to randomize the 'height' of the reflections. The time interval dT between the carrier phrase and the target voice was also varied to determine the effect of the carrier phrase as shown in Fig. 7.1.1.

(2) *Experimental Procedure.* One hundred Japanese Consonant–Vowel (CV) monosyllables were used as the source signals (target voice). The source signals were preceded by a carrier phrase ('*kan kon bai*'). The reverberation sounds produced by this carrier phrase decrease speech intelligibility. The test signals were created by convoluting the sound source signal with computer-generated impulse responses, and then recording them on digital audio tape. Three Japanese test subjects then listened to these test signals, filtered to a range of 0.3–3.4 kHz, through headphones. They had previously been trained for telephone speech intelligibility tests.

D. *Experimental Results*

(1) *Effect of Echogram Patterns.* Figure 7.1.2 shows the results for the speech intelligibility tests. The interval dT was fixed at 200 ms. The articulation loss scores (%) are

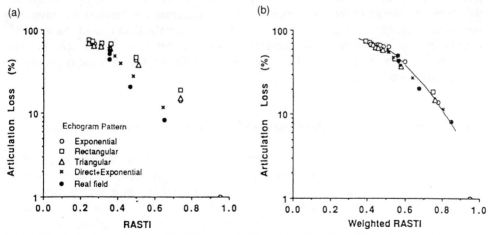

Fig. 7.1.2. Articulation loss vs (a) RASTI and (b) weighted RASTI when dT = 200 ms. From Miyata and Houtgast (1991).

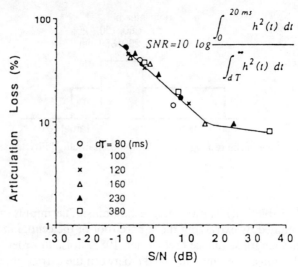

Fig. 7.1.3. Relationship between articulation loss and the signal to noise ratio (S/N). From Miyata and Houtgast (1991).

plotted against both RASTI and weighted RASTI for the various echograms. In Fig. 7.1.2a, which gives the RASTI values, the articulation loss scores for different echograms are scattered. On the other hand, the scores plotted against the weighted RASTI are curvilinear as shown in Fig. 7.1.2b. The weighted RASTI (MTF) method is a good predictor of speech intelligibility for reverberation sound fields.

(2) *Carrier Phrase Effects.* The reverberant sound of the carrier phrase reduces intelligibility. The reverberant energy of the carrier phrase decreases as dT increases. In Fig. 7.1.3, the articulation loss scores are plotted against the signal-to-reverberant sound energy ratio ($=$ SNR, S/N (dB)), as determined by the reverberant sounds of the carrier phrase and the direct and initial part (within 20 ms of the direct sound) of the test word. The articulation loss scores (or speech intelligibility) in reverberant sound fields can be estimated by using S/N. If the total energies of the direct and reverberant sounds are the same, the energy of the initial portion of the echogram can be used for estimating speech intelligibility. In Fig. 7.1.4, S/N is plotted against the RASTI and the weighted RASTI. S/N corresponds to the weighted RASTI better than to the RASTI. This confirms that the Weighted RASTI is a better predictor of speech intelligibility for reverberant sound fields.

7.1.2. Loudspeaker and Microphone Arrays for Improving Speech Intelligibility

Speech intelligibility increases in a reverberant space as the ratio of direct to reverberant sound energy (D/R) increases. Nomura et al. (1991) proposed a loudspeaker (or microphone) array design that improves speech intelligibility in a reverberant space. The D/R at the centre of a spherical surface (or circle) increases as the number of loudspeakers arrayed on the surface increases. The D/R for a sphere has a limit independent of the radius of the surface. This limit is given by the ratio of the equivalent

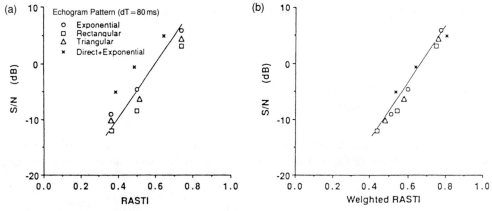

Fig. 7.1.4. S/N vs (a) RASTI and (b) weighted RASTI. From Miyata and Houtgast (1991).

sound absorption area in a room to the squared half-wavelength (uncorrelated minimum distance).

A. D/R for a Single Source in a Reverberant Space

Suppose that there is a sound source in a room. The reverberant energy density in the room is given by

$$R = \frac{4P_0}{Ac} \quad (\text{J/m}^3) \tag{7.1.12}$$

where P_0 is the sound power output of the source, A is the equivalent sound absorbing area (m^2), and c denotes the sound speed (m/s). D/R is given by

$$D/R = \frac{qA}{16\pi r^2} \tag{7.1.13}$$

where r (m) is the distance from the source and q is the directivity factor of the source.

B. D/R due to Multiple Sources in a Reverberant Sound Field

As we discussed in Chapters 3 and 6, the two-point cross correlation coefficients of sound pressure decreases according to $\sin(kr)/kr$, where r is the distance between the two points (m), and k is the wavenumber of the centre frequency of the frequency interval of interest (1/m). Based on the reciprocity (Fahy, 1992) in the sound field, the reverberant sounds from multiple sources are uncorrelated with each other at an observation point if those multiple sources are separated by more than the half-wavelength of the source frequency.

(1) *A Pair of Identical Sources Excited by The Same Signal.* If a pair of identical sources is excited by the same signal, the direct sounds of the two sources are coherent at listening positions located at equal distances from the two sources. If the distance between the pair of sources is larger than the half-wavelength of the source frequency,

and if the sound powers of the two sources are equal, then D/R becomes

$$D/R = \frac{qA}{8\pi r^2} = 2(D/R)_1 \tag{7.1.14}$$

$(D/R)_1$ denotes the D/R for a single source given by equation 7.1.13. The D/R is twice as large as that for a single source.

(2) *Circular Array of Equal-Sound-Power Sources.* Figure 7.1.5 shows an array of sound sources on a circle of radius r. The maximum number of uncorrelated sources for the reverberant sounds is

$$N_{uc} = \frac{4\pi r f}{c} \tag{7.1.15}$$

where f is the source frequency. For this number of uncorrelated sources, the $(D/R)_{cr}$ at the centre of the circle is given by

$$(D/R)_{cr} = \frac{qA}{16\pi r^2} N_{uc} = \frac{qAf}{4rc} \tag{7.1.16}$$

The $(D/R)_{cr}$ becomes N_{uc} times higher than $(D/R)_1$ and, consequently, the $(D/R)_{cr}$ is inversely proportional to the radius of the circle.

(3) *Spherical Array of Sources.* In the case of a spherical array of radius r, the number of uncorrelated sources is given by

$$N_{us} = \frac{16\pi r^2 f^2}{c^2} \tag{7.1.17}$$

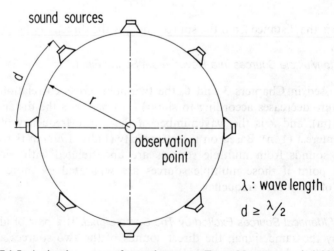

Fig. 7.1.5. A circular array of sound sources. From Nomura et al. (1991).

The $(D/R)_{sp}$ at the center of the sphere therefore becomes

$$(D/R)_{sp} = \frac{qA}{16\pi r^2} N_{us} = \frac{qAf^2}{c^2} \tag{7.1.18}$$

The $(D/R)_{sp}$ is independent of r.

C. Multiple-Layer Arrayed Source and D/R

(1) *Double Coaxial Circular Array.* Figure 7.1.6 shows an example of two concentric circular arrays of sources of radii r_1 and r_2, where $(r_2 - r_1)$ is larger than the half-wavelength. The source sound power is P_2 for sources on the outer circle and P_1 for the inner-layer sources. If the sound sources on the inner circle radiate delayed signals as shown by Fig. 7.1.6, then the $(D/R)_{dcr}$ at the centre becomes (assuming that the number of sources on each circle is $4\pi rf/c$)

$$(D/R)_{dcr} = \frac{(1 + a)^2}{1 + a^2 b} (D/R)_{cr} \tag{7.1.19}$$

$(D/R)_{cr}$ refers to the inner circle, $a = \sqrt{P_2/P_1}$, and $b = r_2/r_1$. $(D/R)_{dcr}$ takes a maximum value when $a = 1/b$

$$(D/R)_{dcr} = \left(1 + \frac{1}{b}\right)(D/R)_{cr} \tag{7.1.20}$$

The sound power of the outer-layer sources must be small (inversely proportional to the circle radius) in order to obtain the maximum D/R. The ratio $(D/R)_{dcr}/(D/R)_{cr}$ is smaller than 2, so the rate of increase in D/R due to the addition of the outer layer decreases as the radius of the outer circle becomes larger.

(2) *Double Coaxial Spherical Array.* The maximum energy ratio at the centre of a double spherical array is twice $(D/R)_{sp}$ when $a = r_1/r_2$. The increase in D/R is independent of the ratio of the radii of the spheres, as long as the difference between the radii is larger

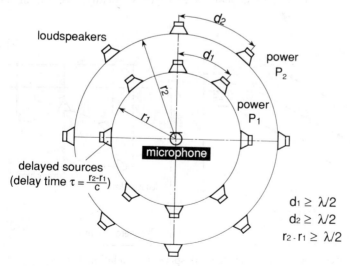

Fig. 7.1.6. Concentric circular loudspeaker arrays. From Nomura et al. (1991).

than the half-wavelength and the maximum number of equally distributed uncorrelated sources at each sphere is $16\pi r^2 f^2/c^2$.

(3) *Limit of D/R in a Room.* The analysis above can be extended to cases of multiple layers of sources. The D/R has a limit imposed by the room acoustic conditions. Equation (7.1.18) can be rewritten as

$$(D/R)_{sp} = \frac{qAf^2}{c^2} = N_{us}(D/R)_{min} \qquad (7.1.21a)$$

and

$$(D/R)_{min} = \frac{q\alpha}{4} \qquad (7.1.21b)$$

where $(D/R)_{min}$ is the energy ratio given by Eq. (7.1.13) when $S = 4\pi r^2$. S is the area of the room surface (m²), α is the average sound absorption coefficient, and N_{us} is the number of uncorrelated points on the room surface.

D. *RASTI Measurement using Loudspeaker Arrays in a Reverberant Field*

We used a large reverberation room (640 m³) with a reverberation time of 28 s at 500 Hz and 11 s at 2000 Hz for these experiments. Figure 7.1.7a shows an example for RASTI versus the number of loudspeakers arrayed on a circle of radius r, as diagrammed in Fig. 7.1.7b. N_{max} is the maximum number of uncorrelated loudspeakers on the circle. The RASTI measured at the centre of the circles increases following the solid lines. We can expect an improvement in speech intelligibility from using circular loudspeaker arrays.

E. *Loudspeaker Linear Array in a Reverberant Field*

Figure 7.1.8 is a schematic diagram of a linear loudspeaker array. This array consists of five omnidirectional type loudspeakers arranged vertically, five amplifiers, and five

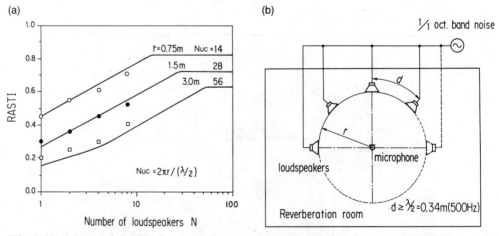

Fig. 7.1.7. Measured RASTI in a reverberant space. (a) Relation between RASTI data and number of loudspeakers. Solid line: theoretical expectation. (b) Configuration of a circular array of loudspeakers. From Nomura et al. (1991).

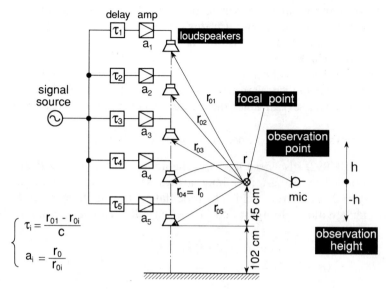

Fig. 7.1.8. Linear array of five loudspeakers with time delays and amplitude adjustments. From Nomura et al. (1991).

delay units. The height of the array is about 3.0 m and the distance between each loudspeaker is 0.45 m (longer than the half-wavelength at 500 Hz). The delay times are set to the values so as to add direct sounds in phase at the focal point. The amplifier gain for each loudspeaker is adjusted to $a_i = r_0/r_i$ in order to obtain the theoretical maximum increase in D/R of about 5 dB at the focal point in a reverberant sound field, assuming point-source characteristics. Figure 7.1.9 shows a plot of the measured RASTI

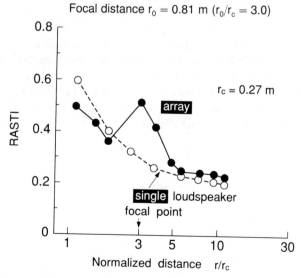

Fig. 7.1.9. RASTI improvement by the linear loudspeaker array. From Nomura et al. (1991).

(as a function of the) normalized distance, for $r_0/r_c = 3.0$, where r_c denotes the critical distance at which D/R becomes unity (in this case, $r_c = 0.27$ m at 500 Hz). This figure illustrates that the maximum RASTI is obtained at the focal point, and that it is nearly 0.2 higher than for a single loudspeaker placed in front of the focal point. These results agree well with the theoretical estimation of a 5 dB increment in the D/R. Moreover, the RASTI for a linear loudspeaker array is better than for a single loudspeaker in a region that extends over several times the critical distance in a reverberant field.

F. *Microphone Arrays for Improving Speech Intelligibility*

Nomura et al. (1992) also described the theoretical increase in D/R and/or D/N that can be obtained by using microphone arrays with time delays and gain adjustments. In noisy and reverberant fields, the modulation transfer function $m(F)$ is given by the following equation (Houtgast and Steeneken, 1985; Houtgast et al., 1980):

$$m(F) = \frac{\left| \int_0^\infty h^2(t) \exp(-j2\pi Ft)\, dt \right|}{\int_0^\infty h^2(t)\, dt} (1 + 10^{K/10})^{-1} \tag{7.1.22}$$

where $h^2(t)$ is the square of the impulse response and

$$K = 10 \log\left[\frac{D}{N}\left(1 + \frac{R}{D}\right)\right] \quad (dB)$$

is the signal-to-noise ratio in dB at the listener's location. If the sound source is an ideal point source (the directivity factor of the source, q, equals 1), the D/R and D/N are

$$D/R = \frac{\alpha S}{8\pi r^2} = 2(D/R)_1 \qquad D/N = \frac{P_0}{2\pi r^2 N_0} = 2(D/N)_1 \tag{7.1.23}$$

for microphones that are equidistant from the source. Here r is the distance from the source (m), α is the average sound absorption coefficient, S is the area of the room (m²), P_0 is the power output of a point source (W), and N_0 is the noise intensity at the observation point (W/m²). $(D/R)_1$ and $(D/N)_1$ denote the D/R and D/N of a single microphone. The D/R and D/N for two microphones is twice as high as for a single microphone. The S/N can be increased by using microphone arrays in a reverberant and noisy fields. Another approach was taken by Kaneda and Ohga (1986). They developed a small-sized microphone array for noise reduction using an adaptation algorithm. Their array technique is quite effective for the reduction of directional noise source signals.

7.1.3. Speech Intelligibility in a Nonexponentially Reverberant Decay Field

The importance of the initial portion of the reverberant energy decay curve has been pointed out for subjective effects (Yegnanarayana and Ramakrishnan, 1975). Nomura et al. (1989) performed speech intelligibility tests in a reverberation room, where the floor was completely covered with absorbing materials, in order to compare speech intelligibility scores with those for exponential decay fields. Speech intelligibility scores can be estimated from the MTF by using the reverberation time of the initial portion of the reverberation energy decay curve.

A. *Nonexponential Reverberation Energy Decay Fields*

Since we can calculate the mean free path in the 3-, 2-, and 1-dimensional fields (Kosten 1960), we apply a superposed decay formula for the three different types of exponential functions (Hirata, 1979; Tohyama and Yoshikawa, 1981). The superposition form is the weighted summation of the group decay functions. The weighting factors are given by the number of modes that are contained in each wave group. The oblique waves become dominant as the frequency increases, since they increase most quickly in density. The relation $n_{pob} > n_{ptan} > n_{pax}$ generally holds well where n_{pob} denotes the modal density of the oblique waves, n_{ptan} the tangential waves, and n_{pax} the axial waves. The sound energy at a steady state is mainly determined by the number of oblique waves. The space-averaged reverberant energy decay is written as

$$E(t) = E_0(t) \frac{1}{n_p} \left\{ \frac{1}{A_{ob}} n_{pob}(\omega) \mathrm{e}^{-A_{ob}ct/4V} + \sum_i \frac{\pi}{4A_{ti}} n_{ptani}(\omega) \mathrm{e}^{-A_{ti}ct/\pi V} + \sum_j \frac{1}{2A_{aj}} n_{paxj}(\omega) \mathrm{e}^{-A_{aj}ct/2V} \right\}$$

$$(7.1.24)$$

where A_i denotes the equivalent sound absorption area (m^2) for the ith group of waves.

Figure 7.1.11 shows an example of a decay curve at 500 Hz (1/1 octave band centre frequency) in a rectangular reverberation room with the floor completely covered with absorbing materials, as shown in Fig. 7.1.10. A loudspeaker was placed 3 m from one corner of the reverberation room, facing the corner, in order to eliminate the influence of the direct sound. Three omnidirectional microphones were placed at three arbitrary positions, but with distance of more than 1 m from the side walls. The initial decay part corresponds to an oblique wave field and the later part corresponds to a tangential wave field. For the oblique wave field T_R is about 3 s and for the tangential wave field, T_R is about 11 s. Figure 7.1.11 shows the frequency characteristics of the reverberation time.

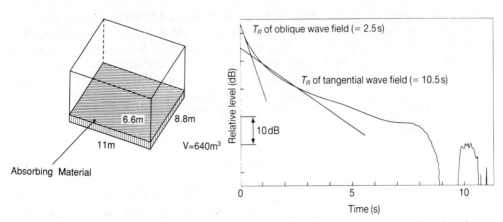

Fig. 7.1.10. Rectangular reverberation room with an absorbing floor and reverberation decay curve at 500 Hz (1/1 octave band centre frequency). From Nomura et al. (1989).

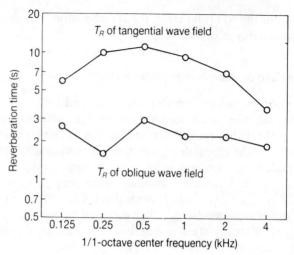

Fig. 7.1.11. Frequency characteristics of the reverberation time with the absorbing floor. From Nomura et al. (1989).

B. *Speech Intelligibility Tests*

Nomura et al. (1989) compared speech intelligibility under two acoustic conditions: with the floor of the reverberation room completely covered with absorbing materials, and with the floor not covered (the empty-room condition). One hundred Japanese CV monosyllables and 50 Dutch Consonant–Vowel–Consonant Phoneme Balanced (CVCPB) words were used in the speech intelligibility tests. The Japanese monosyllables were composed of a carrier phrase ('*kan kon by*'), a consonant, and a vowel. The Dutch words were composed of a carrier phrase, a consonant, a vowel, a consonant, and a carrier phrase. The speech intelligibility tests were made under diotic listening conditions using the following method. The recorded signals were then reproduced from a loudspeaker and recorded using an omnidirectional microphone under both sets of acoustic conditions. The listeners then listened to the recorded test signals through headphones outside the reverberation room. The distance between the sound source and the microphone, which is a very important parameter for speech intelligibility, was 1, 2, 4 and 6 m in the absorbing-floor condition and 0.25, 0.5, 1, 2 and 4 m in the empty-room condition. The loudspeaker was set facing the microphone to simulate a talker speaking to a listener face-to-face. The hearing levels for the listeners were nearly 70 dB (sound pressure level). There were four Japanese listeners and three Dutch listeners. Both groups had previously been specially trained for telephone speech intelligibility tests.

C. *Speech Intelligibility Scores, RASTI, and Reverberation Time*

Speech intelligibility scores are shown in Figs 7.1.12a and b by solid circles for the empty-room and absorbing-floor conditions. They are much higher for the absorbing floor than for the empty room, even when the talker-to-listener distance increases. These results suggest that the T_R from the initial portion of the decay curve governs speech intelligibility. Figure 7.1.12b shows that the RASTI follows the theoretical curve

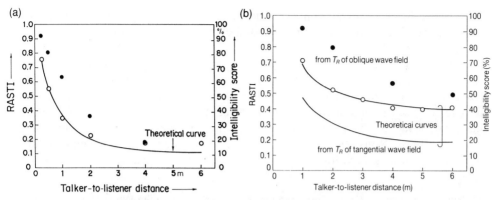

Fig. 7.1.12. Speech intelligibility score (●) and RASTI (○) in the reverberation room: (a) empty-room; (b) absorbing-floor. From Nomura et al. (1989).

(solid line) obtained using the T_R of the initial part of the decay curve. This result demonstrates that speech intelligibility scores are related to the T_R of the initial part of the reverberant energy decay process. Nonexponential decay fields occur in a large room with an absorbing floor and parallel reflective walls. An absorbent floor can improve the speech intelligibility, although the reverberation time is long. We expect that an acoustic space can be designed to have high intelligibility with long reverberation.

7.2. Sound Source Localization in a Reverberant Space

A human listener can detect a sound source in a reverberant environment. This is interpreted in terms of the temporal changes in interaural cross correlation and precedence effects. This chapter describes the temporal aspects of the interaural cross correlation. The transient interaural cross correlation is described following Blauert (1983) and Yanagawa (1989).

7.2.1. Interaural Cross Correlation for Nonstationary Signals

An integrated evaluation of subjective impressions of sound fields created by music or other transient signals requires an understanding of how the interaural cross correlation changes over time (Mori et al., 1987). When a white noise signal $s(t)$ is switched on at $t = 0$ and off at $t = T$, the temporal change in cross correlation function between two observation locations (both ears) is given by

$$\Phi_{12}(t, \tau) = \left\langle \int_0^t s(u) h_1(t - u)\, du \int_0^t s(u') h_2(t + \tau - u')\, du' \right\rangle \qquad (7.2.1)$$

where $\langle\ \rangle$ denotes taking an ensemble average, and $h_1(t)$, $h_2(t)$ are impulse responses from the source to the two observation locations. Suppose that the signal length T is sufficiently longer than the length T_H of the impulse responses (almost equal to the reverberation time, T_R). The above equation shows the temporal changes in cross correlation function during the build-up process, when $t < T_H$. As t is larger than T_H (but smaller than T), the cross correlation function is the same as in the steady state, which is determined by the time lag τ. Finally, in the case that $T < t < T + T_H$, Eq.

(7.2.1) gives the cross correlation in the decay process. If the signal length T is shorter than T_H, then Eq. (7.2.1) shows the temporal changes in the cross correlation function excited by burst-like noise signals.

Figure 7.2.1 illustrates the temporal changes in cross correlation function in the build-up process in a room (volume 100 m³, reverberation time 0.3 s) (Yanagawa et al., 1988). The maximum value of the cross correlation function decreases as time passes. We can surmise that a human listener can detect the sound source locations, since the maximum value of the interaural cross correlation function is nearly unity in the build-up process. Nonstationary signals such as music or speech involve many build-up and decay processes, such high correlation values occur frequently.

A. Transient Interaural Cross Correlation in Relation to Intermittent White Noise

Speech signals have many silent gaps. We are interested in the temporal changes of the interaural correlation due to the silent gaps and their effects on hearing. We describe the temporal changes using a simple model of the time gaps following Yanagawa (1989).

(1) *Interaural Cross Correlation during the Time Gap and Definition of Transient Interaural Cross Correlation.* Suppose that the impulse responses from the sound source to the two ears are labelled $h_1(t)$ and $h_2(t)$. The interaural cross correlation function for the

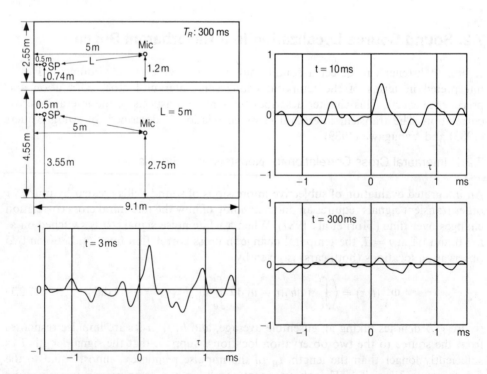

Fig. 7.2.1. Temporal changes of cross correlation function between two points during a build-up process in a reverberant space (distance between two microphones = 33 cm). After Yanagawa et al. (1988).

build-up process of a white noise signal is described by

$$\Phi_{up12}(t_{up}, \tau) = \left\langle \int_0^{t_{up}} s(u)h_1(t-u)\, du \int_0^{t_{up}} s(u')h_2(t+\tau-u')\, du' \right\rangle$$

$$= \int_0^{t_{up}} h_1(t)h_2(t+\tau)\, dt \tag{7.2.2}$$

The normalized cross correlation function is given by

$$R_{up12}(t_{up}, \tau) = \frac{\displaystyle\int_0^{t_{up}} h_1(t)h_2(t+\tau)\, dt}{\left(\displaystyle\int_0^{t_{up}} h_1^2(t)\, dt\right)^{1/2} \left(\displaystyle\int_0^{t_{up}} h_2^2(t)\, dt\right)^{1/2}} \tag{7.2.3}$$

where $t = 0$ is the start time of the signal. Similarly, in the decay process after the signal stops, we have

$$R_{down12}(t_{down}, \tau) = \frac{\displaystyle\int_{t_{down}}^{\infty} h_1(t)h_2(t+\tau)\, dt}{\left(\displaystyle\int_{t_{down}}^{\infty} h_1^2(t)\, dt\right)^{1/2} \left(\displaystyle\int_{t_{down}}^{\infty} h_2^2(t)\, dt\right)^{1/2}} \tag{7.2.4}$$

where t_{down} is elapsed time after the signal stopped. Houtgast (1988) suggested that the temporal change of the interaural cross correlation during the silent portion is important for sound image localization in reverberation. Yanagawa (1989) analysed such temporal changes, introducing the transient interaural cross correlation (TRICC) value.

Now we shall focus on the silent portion of the intermittent white noise, and on the response of the sound field to the build-up of the continuous signal that follows this silent portion. If the silent portion is shorter than the reverberation time, the reverberant sound will overlap the build-up process of the following sound. During this period of overlap, the interaural cross correlation functions are

$$R_g(t_{up}, \tau) = \frac{\sqrt{E_{up1}E_{up2}}\, R_{up}(t_{up}, \tau) + \sqrt{E_{down1}E_{down2}}\, R_{down}(t_{up}+T_g, \tau)}{\sqrt{E_{up1}+E_{down1}}\sqrt{E_{up2}+E_{down2}}} \tag{7.2.5}$$

where T_g is the length of the silent portion, and

$$E_{up1} = \int_0^{t_{up}} h_1^2(t)\, dt, \qquad E_{up2} = \int_0^{t_{up}} h_2^2(t)\, dt$$

$$\tag{7.2.6}$$

$$E_{down1} = \int_{t_{up}+T_g}^{\infty} h_1^2(t)\, dt, \qquad E_{down2} = \int_{t_{up}+T_g}^{\infty} h_2^2(t)\, dt$$

Similarly to the definition of IACC, we call the maximal value of R defined above for $|\tau| \leqslant 1$ ms the TRICC (transient interaural cross correlation).

(2) *Experimental Results.* Typical examples of TRICC are shown in Fig. 7.2.2. The horizontal axis (x-axis) shows elapsed time from $t_{down} = 0$. In these results the values for energies, E_{up} and E_{down}, are replaced by exponential functions containing time constants determined by reverberation time. The curves show the temporal changes in TRICC with T_g as a parameter. Figure 7.2.2a shows the results for a room with

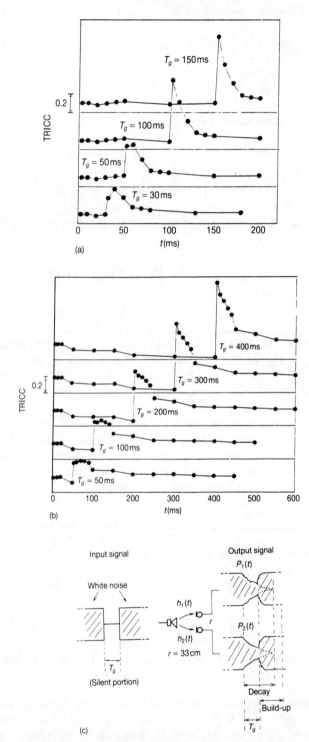

Fig. 7.2.2. Temporal changes of TRICC when the signal contains a single silent portion (length of silent portion = T_g): (a) room; (b) auditorium; (c) signal configuration. After Yanagawa (1989).

reverberation time of 0.3 s, while Fig. 7.2.2b is for an auditorium (volume 5240 m³) with reverberation time of 1.3 s. These values of TRICC increase during the build-up of the second continuous sound after the silent portion (gap) (see Fig. 7.2.2c). However, the shorter the time gap, the greater is the energy of the reverberant sound. The decay process overlapped with the build-up portion. A shorter time gap (silent portion) means a decrease in TRICC in the build-up process, with values approaching those of the steady state. Figure 7.2.3 plots the maximal values of TRICC in the build-up process shown in Fig. 7.2.2. These results make clear the linearity of changes in relation to length of the silent portion T_g, and show how the values gradually approach those of the steady state. Subjective diffuseness occurs in the decay process, where TRICC is small. In the next build-up process, TRICC shows maximal values, and the sound image is located as being in the direction of the sound source. The decrease in TRICC maximal values with decreases in T_g indicates that with this build-up process in the sound field the sound becomes progressively more reverberant and the sound image becomes harder to locate.

B. *Perception of Variation in Subjective Diffuseness due to Temporal Changes in IACC*

Mori et al. (1987) investigated the changes in subjective diffuseness when IACC is changed during stimulation. The stimulus is 1/3-octave band noise with a centre frequency of 500 Hz. In an anechoic room, one loudspeaker was placed facing the subjects, and two others at left and right angles of 54°. Sound was presented first from the centre loudspeaker only, with an IACC of 1.0. Then stimuli that were incoherent in relation to the centre loudspeaker stimulus were presented from the left and right loudspeakers, with a delay of Δt and IACC of 0.3. The six delay times Δt used in their experiment were 0, 12.5, 25, 50, 100 and 200 ms. The duration of each stimulus was 400 ms, with a pause of 600 ms between stimuli. The A-weighted sound pressure level at which stimuli were given was 74 dB at the centre of the head of the subject. Twelve students were used as subjects. One of the six stimuli above was presented five times, then the subject was asked to determine whether there was a change in subjective diffuseness during the 10 s immediately following the stimulus. The five-burst stimuli

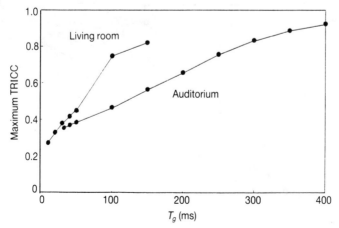

Fig. 7.2.3. Maximum value of TRICC at the build-up process versus length of silent portion. After Yanagawa (1989).

were presented in random order. The study determined the success rate with which each subject was able to perceive the change in subjective diffuseness for each stimulus. The experimental results are shown in Fig. 7.2.4. There was a great deal of variation from one subject to another, but some of the subjects were actually able to perceive change over a time span of less than 50 ms. We will discuss the perception of the change of IACC in the next section in relation to the precedence effect.

7.2.2. Precedence Effect in a Reverberant Space

The precedence effect (Haas, 1951, 1972) originally was applied to the situation in which two or more mutually delayed versions of a sound (typically speech or music) reach a listener from different directions. Under certain conditions of relative delays and amplitudes, the total sound is localized according to the direction of the sound that first reaches the listener. The effect plays an important role in sound localization under reverberation condition. The precedence effect is also manifested when stimuli are presented by headphone. In that case it refers to the perception of lateralization, governed by the dichotic cues interaural time delay (IATD) and interaural level difference (IALD). For instance, when two binaural impulses with different IATDs are presented within a few milliseconds, this pair of impulses is perceived as one sound with a lateral position dominated by the IATD at the first pulse. In case of fluctuating sound, such as a music signal in a concert hall, the interaural cross correlation is not constant over time. At the onset of a sound, the correlation is high (dominated by the direct sound), and decreases to a lower value associated with the reverberant sound, as shown in Fig. 7.2.1. In case of such a changing interaural correlation, a precedence effect will have its effect on the binaural perception. Aoki and Houtgast (1992) compared the traditional precedence effect in the perception of lateralization (based on the dichotic cue IATD) and a possibly similar effect in the perception of the blurred or compact sound image (based on the interaural phase relation, IAPD, but for only two cases:

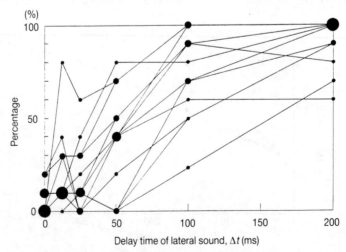

Fig. 7.2.4. Percentage perception of the change of subjective diffuseness as a function of the delay time of lateral sound. (The size of a circle denotes the number of subjects who responded.) From Mori et al. (1987).

in-phase or antiphase). For that purpose they adopted an experimental paradigm that allows for a quantitative assessment of the precedence effect, both with respect to lateralization and with respect to the perception of sound image compactness.

A. *Test Signals*

The test signal for one trial consists of two blocks (Block A and Block B in Fig. 7.2.5). Each block is subdivided into two parts, a leading part (T_L) and a trailing part (T_t), and the two parts contain opposite information on either IATD or IAPD. In one trial

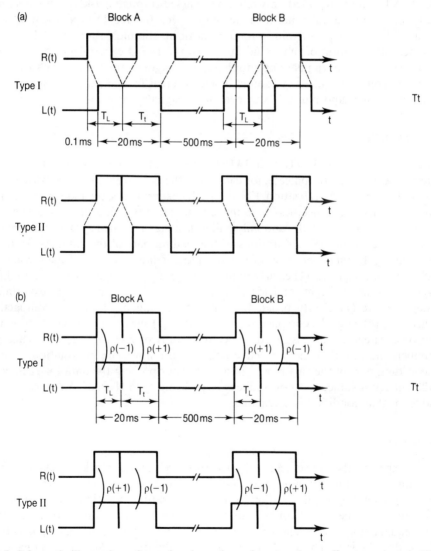

Fig. 7.2.5. Schematic illustration of test signal envelopes for precedence effect experiments. The carrier is 7 kHz low-passed white noise. The right-ear signal is indicated by $R(t)$, the left-ear signal by $L(t)$. (a) Interaural time difference of plus or minus 0.1 ms. (b) Interaural phase difference of 0 or 180°. For type I and type II, the order of presentation of the two blocks within a trial is reversed. From Aoki and Houtgast (1992).

with the two blocks, the information in the second block is reversed compared to that in the first block. In type I of Fig. 7.2.5a the leading and trailing parts in block A have information on right and left lateralization, respectively (opposite values of IATD). The order is reversed in block B. Two types of trials were prepared so as to cancel any effect of response bias associated with the order of presentation of the two blocks. The type I trial contains block A followed by block B, and in the type II trial the order is reversed. Two parameters were used in varying the relative weight of the leading and trailing part of the stimulus. The point of separation between the two parts was varied in 2-ms steps from 0 to 20 ms with a constant total duration of 20 ms. The amplitude ratio of the leading and trailing part could be set to -18, -12, -6, -3, 0, 6, 12 or 18 dB. All stimuli are based on white noise passed through a 7-kHz low-pass filter. The value of the interaural time difference chosen for the IATD experiment was plus or minus 0.1 ms. For an ongoing broad-band noise stimulus, such an IATD leads to a distinct lateralization of the sound image. In the IAPD experiment, the signals at the two ears were either in phase or antiphase. For ongoing noise, the in-phase signal leads to a very compact in-the-head sound image, whereas for the antiphase signal the image is much broader and blurred. The level of the signals was set to approximately 70 dB.

B. *Test Method*

The experiments on IATD and IAPD were conducted separately. For IATD, the subjects were asked to indicate, for each trial, the perceived difference between the two blocks in terms of lateralization. For the IAPD experiments, the judgements were based on the quality of compactness/broadness. Four subjects judged five test sets each. A set consisted of 66 trials with randomly edited test signals, three type I and three type II signals for each of the 11 durations of the leading part of the stimulus. In the IATD experiment, the image position of the first block relative to that of the second block is rated in five categories: (1) clearly right to left, (2) vaguely right to left, (3) no difference, (4) vaguely left to right, and (5) clearly left to right. In the IAPD experiment, the categories are (1) clearly broad to compact, (2) vaguely broad to compact, (3) no difference, (4) vaguely compact to broad, and (5) clearly compact to broad. These response categories are assigned respective scores of 2, 1, 0, -1, -2. Thus positive numbers indicate that the subjective impression follows the information presented in the leading part of the two blocks in a trial. After pooling the data for each condition (120 entries: 4 subjects, 3 noise tokens, type I and type II trials, 5 replicas), the mean value for that condition is calculated.

C. *Results*

In all experiments, the subjects reported to perceive each block (block A or block B) as a single sound image in terms of lateralization or compactness. This is a prerequisite for the present approach based on the overall sensation of a stimulus with time-varying dichotic cues. The results for the condition in which the leading and the trailing part have equal amplitude are shown in Fig. 7.2.6. The duration of the leading part is plotted along the abscissa. The numbers along the ordinate indicate the dominance of the leading part in the subject's judgements. For instance, '2' indicates that the judgement is strongly dominated by the information in the leading part, and '0' indicates that the influence of the leading part is balanced by that in the trailing part of the stimulus.

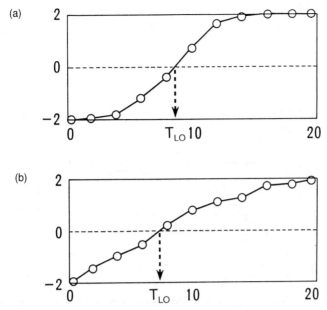

Fig. 7.2.6. The results with equal amplitudes of leading and the trailing parts: (a) IATD results; (b) IAPD results. Abscissa: duration of leading part T_L in Fig. 7.2.5 (leading plus trailing part is 20 ms). Ordinate: subjective impression, where positive values indicate a dominance of the leading part over the trailing part. From Aoki and Houtgast (1992).

For both conditions (IATD and IAPD) the data indicate that for equal physical weight ($T_L = T_t = 10$ ms), the overall impression is slightly dominated by the information in the leading part. Thus, the manifestation of the precedence effect for the present type of stimulus is only small. However, it is significant that the results are very similar for the two types of dichotic information, IATD and IAPD. The dominance of the leading part may be balanced by shortening its duration. The duration required for an equal weight of the two parts in the overall sensation is taken as a quantitative measure for the precedence effect, and is determined by linear interpolation as illustrated in Fig. 7.2.6: the value of T_{LO} that is associated with an average overall judgement '0'. This value, both for IATD and IAPD, is plotted in Fig. 7.2.8 for the condition of 0 dB level difference between leading and trailing parts.

Similar data were obtained for a range of values of the level difference between the two parts of the stimulus. (As indicated before, the values ranged between -18 dB and 18 dB.) As an example, Fig. 7.2.7 presents the results for the condition where the level is 12 dB higher for the leading part than for the trailing part. In that case, the dominance of the information in the leading part is stronger: T_L must be reduced to under 5 ms to compensate for the dominance of the information in the leading part over that in the trailing part. The actual T_{LO} values associated with an average overall judgement of '0', derived by linear interpolation as illustrated in Fig. 7.2.7, are plotted in Fig. 7.2.8 for the -12 dB level difference. The other entries in Fig. 7.2.8 were derived in a similar way. Figure 7.2.8 shows the trade-off relationships between the duration of the leading stimulus and the level differences, under the condition that subjects do not perceive a systematic difference between the sound images of the first and the second

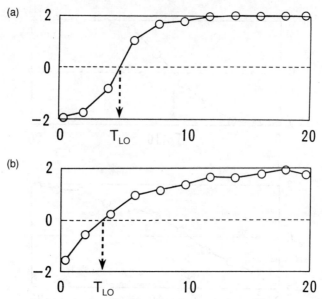

Fig. 7.2.7. The results with the amplitudes of leading and the trailing parts differing $-12\,\mathrm{dB}$: (a) IATD results; (b) IAPD results. Abscissa and ordinate as Fig. 7.2.6. From Aoki and Houtgast (1992).

block. In other words, Fig. 7.2.8 displays the conditions for which the precedence effect is compensated for by shortening the duration and/or lowering the level of the leading part relative to the trailing part. The amount of compensation required in the case of IAPD is very similar to that required in the case of IATD. The precedence effect for both types of information is essentially the same. The dashed lines in Fig. 7.2.8 show

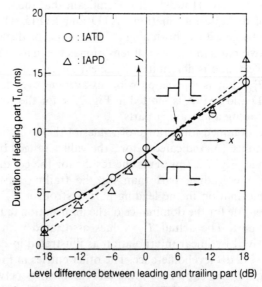

Fig. 7.2.8. Illustration of precedence effect on both IATD and IAPD. From Aoki and Houtgast (1992).

linear regressions by the least-squares method. The equations are

IATD:

$$y = 0.32x + 8.16$$

IAPD:

$$\text{(7.2.7)}$$

$$y = 0.38x + 7.93$$

These equations are almost identical. When there is no level difference between leading and trailing parts, the duration of the leading part should be about 2/3 the duration of the trailing part in order to compensate for the precedence effect (8 and 12 ms, respectively). When the durations of leading and trailing parts are the same (10 ms), the precedence effect can be compensated for by reducing the level of the leading part by about 5.5 dB. A level reduction of 5.5 dB corresponds to a loudness reduction by a factor of 0.68 when applying the relation (Stevens, 1957),

$$L = cI^{0.3} \tag{7.2.8}$$

where L is loudness, I is intensity and c is a constant. That is, $(10^{-5.5/10})^{0.3} = 0.68$. The factor 0.68 is very close to the above-mentioned 2/3. The weight of information is affected equally by duration and by the quantity $I^{0.3}$.

The data points in Fig. 7.2.8 referring to the condition where the leading and the trailing parts are equally effective are assumed to correspond to the condition

$$WI_L^{0.3} T_L = I_t^{0.3} T_t \tag{7.2.9}$$

Here, I_L and I_t denote the intensities of the leading and trailing parts, and T_L and T_t are their respective durations. The weighting factor W is introduced to quantify the precedence effect: $W > 1$ indicates that the leading part has a stronger effect than the trailing part. Equation (7.2.9) can be applied to calculate the value of W for each individual data point in Fig. 7.2.8. For each data point, the level difference (abscissa) determines the ratio I_L/I_t, and the duration (ordinate) determines the ratio T_L/T_t. The results are presented in Table 7.2.1. All data points lead to about the same value of W (excluding the data for the -18 dB level difference), indicating that with $W = 1.5$, Eq.

Table 7.2.1. Derivation of weighting factor.
From Aoki and Houtgast (1992).

Level difference (dB)	W (calculated)	
	IATD	IACC
(−18)	(3.11)	(3.32)
−12	1.42	1.80
−6	1.51	1.93
−3	1.25	1.61
0	1.33	1.67
+6	1.61	1.64
+12	1.69	1.59
+18	1.45	0.84
Mean	1.46	1.58
S.D.	0.14	0.32

(7.2.9) does indeed account for the trade-off between differences in duration and/or level to compensate for the precedence effect. This holds equally well for the IATD and the IAPD data. The solid curve in Fig. 7.2.8 is drawn according to Eq. (7.2.9) with $W = 1.5$. The solid curve is close to the dashed lines derived from Eq. (7.2.7). For small values of the level difference between the trailing and leading part $(L_t - L_L)$, Eq. (7.2.9) can be approximated by

$$T_L = 0.069 T_a (L_t - L_L) \frac{W}{(1 + W)^2} + \frac{T_a}{(1 + W)} \tag{7.2.10}$$

$$= 0.33(L_t - L_L) + 8$$

where $T_a = T_L + T_t = 20$ ms and $W = 1.5$. Equation (7.2.10), being the first-order linear approximation of the relation between T_L and $(L_t - L_L)$ as given by Eq. (7.2.9), represents a line that is very similar to the regression lines given by Eq. (7.2.7).

7.3. Sound Field Control for Concert Halls

This section describes digital signal processing technologies for the concert hall acoustics. Sound transmission measurement and sound field equalizing filters are particularly important for diagnosing and controlling a sound field in a concert hall.

7.3.1. Time-spatial Sound Transmission Measurement and Visualization

An impulse response record is a time history of the reflections reaching a reception point. Not only is the impulse response record important, but obtaining time-spatial records is also important.

A. *Closely Located Four-Point Microphone Method (Yamasaki and Itow, 1989)*

Suppose that a sampled impulse response record is obtained at a listening point. Each observed datum in the response can be representative of a reflected sound that is coming to the listening point from a virtual source. We can identify the position of the virtual sources in a 3-dimensional space if we obtain the impulse responses using four microphones that are located close to each other at the listening point. Figure 7.3.1 illustrates this scheme where the observing points are located on the rectangular coordinate axis and at the origin. We define r_{on}, r_{xn}, r_{yn} and r_{zn} as the distances between a virtual source N and each of the observation points X, Y and Z, and the origin O. The coordinate of a virtual source N for each reflection (X_n, Y_n, Z_n) is

$$X_n = \frac{(d^2 + r_{on}^2 - r_{xn}^2)}{2d}$$

$$Y_n = \frac{(d^2 + r_{on}^2 - r_{yn}^2)}{2d} \tag{7.3.1}$$

$$Z_n = \frac{(d^2 + r_{on}^2 - r_{zn}^2)}{2d}.$$

Figure 7.3.2 shows a model and the impulse responses at each observation point.

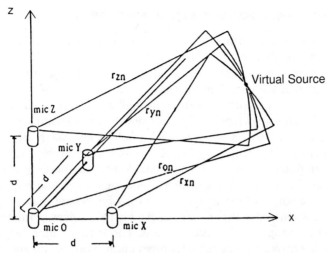

Fig. 7.3.1. A virtual-source location. The location of a virtual source is the intersection of four sphere surfaces. The radii are r_{on}, r_{xn}, r_{yn} and r_{zn}, respectively. From Yanagawa et al. (1988).

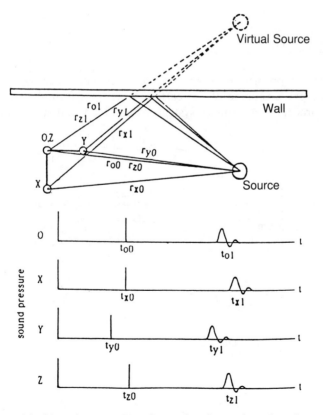

Fig. 7.3.2. A model of impulse responses observed at four-point microphones. From Yanagawa et al. (1988).

B. *Measurement of Virtual Sources in a Large Lecture Hall*

The virtual source measurements were made in a lecture hall with a volume of 5420 m^3 and a reverberation time of 1.3 s. The sound source consisted of a pair of 20-cm diameter full-range loudspeakers without cabinets, attached face to face. This sound source was located at the centre of the stage at a height of 1.5 m, while the four sound reception points were at the front centre of the hall at a height of 1.2 m above the floor. The distance of each sound reception point from the sound source was 7.7 m (see Fig. 7.3.3). The input signal fed in-phase to each loudspeaker was a series of rectangular pulses at nonperiodic intervals. No interval was shorter than the reverberation time of the hall. The amplitude of the pulses was 100 V and the pulse width was 5 μs. The periodic undesired noise was reduced after averaging the impulse responses.

The temporal changes of the virtual source distribution are shown in Figs 7.3.4 and 7.3.5. Figure 7.3.4 illustrates the changes of the virtual sources in the build-up process and Fig. 7.3.5 the changes in the decay process. The virtual source distributions shown in both figures are projections of virtual sources on the floor or side surfaces. The radius of the circle represents the power of the virtual source. The range of the distribution shown in the build-up process is approximately 200 ms from the time of direct sound arrival. The number of virtual sources increases with elapsed time. This method developed by Yamasaki and Itow (1989) has been used to obtain the time-spatial records for concert halls.

7.3.2. Howling Cancellers

Howling is a common problem in room acoustics. Schroeder's (1964) or Waterhouse's (1965) papers are perhaps some of the earliest work that formulated the howlback phenomenon theoretically in terms of system instability. The first step in analysing a

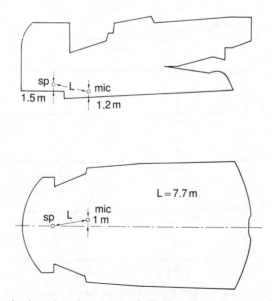

Fig. 7.3.3. Receiving point locations in a lecture hall. From Yanagawa et al. (1988).

(a)

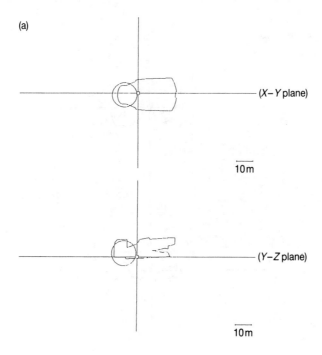

(X−Y plane)

10m

(Y−Z plane)

10m

(b)

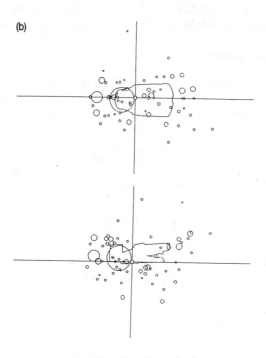

Fig. 7.3.4 (*Contd overleaf*)

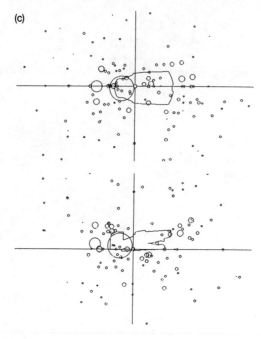

Fig. 7.3.4. Virtual sources in a build-up process: (a) $t = 0$–3 ms; (b) $t = 0$–100 ms; (c) $t = 0$–500 ms. From Yanagawa et al. (1988).

combined electroacoustic system (Fig. 7.3.6) is to consider an electrical amplifier with feedback. The criterion for stability has been given by Nyquist (1932).

A. Nyquist Criterion for Stability

For Nyquist's feedback system both the amplifier gain μ and the gain or loss β of the feedback signal are complex quantities in order to account for changes in amplitude and the phase of the signal. The closed-loop transfer function (TF) can be written as

$$G(\omega) = \frac{\mu}{1 - \mu\beta} \tag{7.3.2}$$

We call a system stable when

$$\int_0^\infty |g(t)| \, dt < \infty \tag{7.3.3}$$

where $g(t)$ is the impulse response of the closed-loop TF G. The system is unconditionally stable at all frequencies, except for those where two conditions are met:

$$\text{Im}(\mu\beta) = 0, \quad \text{Re}(\mu\beta) \geqslant 1 \tag{7.3.4}$$

These conditions are equivalent to

$$\theta = 2n\pi, \quad n = 0, 1, 2, \ldots, \quad |\mu\beta| \geqslant 1 \tag{7.3.5}$$

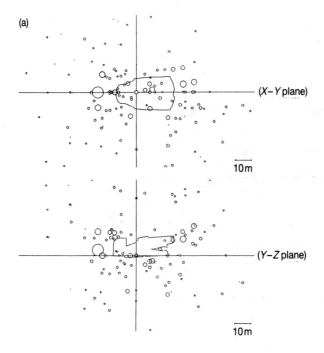

(a)

(X–Y plane)

10m

(Y–Z plane)

10m

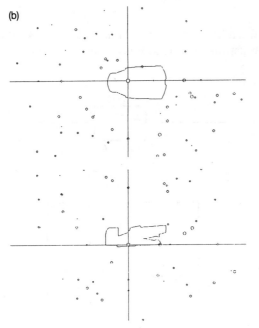

(b)

Fig. 7.3.5. Virtual sources in a decay process: (a) $t = 3$–500 ms; (b) $t = 100$–500 ms. From Yanagawa et al. (1988).

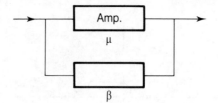

Fig. 7.3.6. Model of a feedback system. After Waterhouse (1965).

B. *Howling Control using an Inverse Filter*

Now, let us consider a feedback system that includes the room TF (Figure 7.3.7),

$$G(\omega) = \frac{H(\omega)}{1 - AH(\omega)} \tag{7.3.6}$$

where $H(\omega)$ is the room TF in the space. The room TF H can be written as

$$H = |H|\, e^{-j\theta_{\text{Min}}}\, e^{-j\theta_{\text{All-pass}}} \tag{7.3.7}$$

The TF in a closed loop can be rewritten as

$$G(\omega) = \frac{H}{1 - AH} = \frac{|H|\, e^{-j\theta_{\text{Min}}}\, e^{-j\theta_{\text{All-pass}}}}{1 - A|H|\, e^{-j\theta_{\text{Min}}}\, e^{-j\theta_{\text{All-pass}}}} \tag{7.3.8}$$

An inverse filter is available for the minimum-phase component. If we insert this inverse filter into the feedback loop as shown in Fig. 7.3.8, the TF becomes

$$G(\omega) = \frac{e^{-j\theta_{\text{All-pass}}}}{1 - A e^{-j\theta_{\text{All-pass}}}}. \tag{7.3.9}$$

This system is stable when $|A| < 1$. The minimum-phase inverse filter is sensitive to the variations of TFs. A cepstrum-smoothed inverse filter is robust for inverse processing.

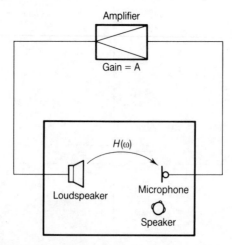

Fig. 7.3.7. Feedback system including the room transfer function. $H(\omega)$ = transfer function (TF) from the loudspeaker to the microphone.

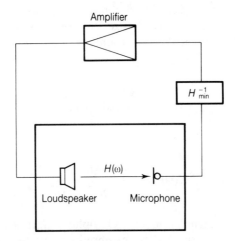

Fig. 7.3.8. Minimum-phase inverse filter and a feedback system including the room transfer function. From Ushiyama et al. (1994).

Figure 7.3.9 shows an example of inverse filtering by a smoothed inverse filter (Ushiyama et al., 1994). We can expect increase of the howling margin by such a robust inverse filter.

C. *Adaptive Control for a Feedback System*

Imai et al. (1992) are developing a stable public address system that cancels the feedback signal by using an adaptive filter such as an echo canceller. The schematic

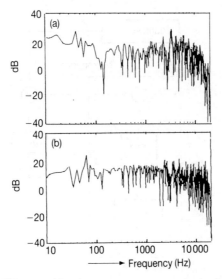

Fig. 7.3.9. An example of TF smoothing by a cepstrum-smoothed minimum-phase inverse filter (a) without inverse filter, (b) with inverse filter. From Ushiyama et al. (1994).

diagram is shown in Fig. 7.3.10. This system is composed of an adaptive filter and a delay circuit. The adaptive filter synthesizes the feedback signal into the microphone from the loudspeaker through the room transfer function. The input signal to the microphone in the closed loop is written as

$$x(t) = s(t) + \{h(t) - \widehat{h(t)}\}*x(t) \qquad (7.3.10)$$

In the frequency plane, we have

$$X(\omega) = \frac{S(\omega)}{1 - \{H(\omega) - \widehat{H(\omega)}\}} \qquad (7.3.11)$$

where $s(t)$ is the direct signal into the microphone, $h(t)$ denotes the room impulse response, $\widehat{h(t)}$ means the estimated room impulse response using the adaptive filter, $S(\omega)$, $X(\omega)$ are the Fourier transforms, and $H(\omega)$, $\widehat{H(\omega)}$ are the room TF and the estimated TF, respectively. If we assume $H(\omega) \approx \widehat{H(\omega)}$, then $X(\omega) \approx S(\omega)$. The closed loop is thereby cut out and consequently we have eliminated any possibility of singing.

We, however, cannot observe the error signal due to $H(\omega) - \widehat{H(\omega)}$ since the microphone input signal includes both the direct signal from the talker and the feedback signal from the loudspeaker. The adaptive process will cancel both the direct signal and the reproduced signal from the loudspeaker. The audience will not be able to hear anything from the talker.

The delay unit in Fig. 7.3.10 makes it possible to preserve the direct sound. The amount of delay time is determined by the correlation function of the signal source. The input signal can be written as

$$x(t) = s(t) + \{h(t) - \widehat{h(t)}\}*x(t) = s(t) + \varepsilon_h(t) \qquad (7.3.12)$$

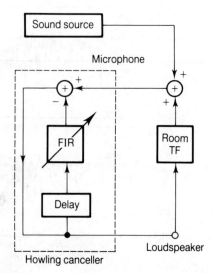

Fig. 7.3.10. Schematic diagram of a howling canceller. From Imai et al. (1992).

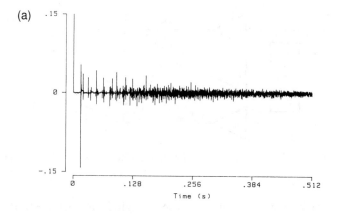

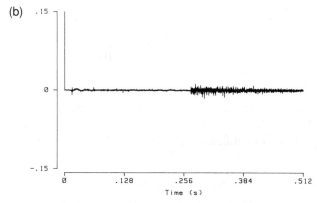

Fig. 7.3.11. Impulse response of a closed-loop system including a room TF simulated by a reverberator: (a) without the howling canceller; (b) with the canceller. From Imai et al. (1992).

where $\varepsilon_h(t)$ denotes the error signal due to the estimation error of the room TF. This error signal should be reduced to zero. If there is no correlation between the direct sounds $s(t)$ and the feedback signal, then the expected power of the input signal can be expressed as

$$\langle x^2(t) \rangle = \langle s^2(t) \rangle + \langle \varepsilon_h^2(t) \rangle \qquad (7.3.13)$$

Consequently, if the direct sound power is constant, the expected total power also converges to its minimum value, as $\langle \varepsilon_h^2(t) \rangle$ becomes minimum. Figure 7.3.11 illustrates a sample of the simulations by Imai et al. (1992). An artificial reverberator was used as a model of the TF. The TF was estimated by the adaptation scheme using a white noise signal.

7.3.3. Digital Equalizing Filters for Transfer Functions

We introduce a digital equalizer for equalizing undesired frequency responses of transfer functions in a concert hall (Hirata 1981, Iizuka et al. 1991). Figure 7.3.12

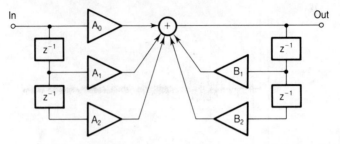

Fig. 7.3.12. A second-order IIR digital filter. From Hirata (1981).

illustrates the block diagram of an IIR type of digital filter whose transfer function is written as

$$H(z) = \frac{A_0 + A_1 z^{-1} + A_2 z^{-2}}{1 - B_1 z^{-1} - B_2 z^{-2}} \qquad (7.3.14)$$

where $z^{-1} = e^{-j2\pi f/F}$, f is the frequency (Hz), and F is the sampling frequency (Hz). In designing an audio equalizer, it is desirable to control the characteristics using the parameters such as the centre frequency (f_0), the bandwidth and the gain. To achieve this, Iizuka et al. (1991) modified the $H(z)$ to

$$H(z) \approx \frac{\dfrac{f_0}{f} - \dfrac{f}{f_0} + j2\alpha R}{\dfrac{f_0}{f} - \dfrac{f}{f_0} + j2\beta R} \qquad (f_0 \ll F) \qquad (7.3.15)$$

where $R = 1/2Q$ and $G = 20\log(\alpha/\beta)$.

The equalizing response has a peak at f_0. If we let $\beta = 1$ and define $\alpha > 1$, the height of the peak is given by α. For a valley, if we let $\alpha = 1$ and $\beta > 1$, the depth of the valley is given by β. A notch filter is attainable by setting $\alpha = 0$ and $\beta = 1$. Figure 7.3.13 illustrates examples of the equalizer response and the notch filter response.

7.4. A Design Study for Concert Halls

The theory of subjective preference and the related model of the auditory–brain system (Chapter 5) may be used in designing the acoustics of a concert hall. The global subjective preference at each seat may be obtained by calculating the impulse response and four acoustic factors for the concert hall. The design problem is known as an inverse problem, so there are many possible shapes for the hall. To demonstrate this, one design procedure is shown here. Because the preference greatly depends on the individual, a seat selection system is recommended to be introduced after the construction of a concert hall.

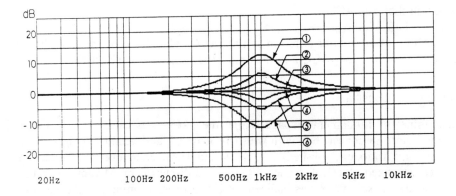

$$f_o = 1000 \ [Hz]$$
$$Q = \ 3$$

①   G = +12 [dB]
②   G = + 6 [dB]
③   G = + 3 [dB]
④   G = - 3 [dB]
⑤   G = - 6 [dB]
⑥   G = - 12 [dB]

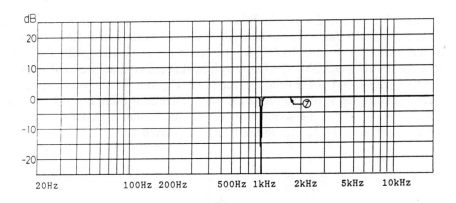

⑦ $f_o = 1000 \ [Hz]$, $Q = 30$, $G = -50$ [dB]

Fig. 7.3.13. Examples of the equalizers. From Hirata (1981).

7.4.1. Control of Temporal Criteria

The first step in this design procedure is to establish the range of source signals and the values of the effective duration of ACF according to the purpose of the concert hall. Here we assume that we want values suitable for chamber music, say, $\tau_e = 40$–85 ms. Then the reverberation time may be determined by maximizing subjective preference in the temporal criteria. The most preferred reverberation time obtained is centred on $1.5 \ s \approx 23 \times 65$ ms (occupied). Together with the audience capacity, the dimensions of the hall related to Δt_1 may be determined.

7.4.2. Control of the IACC

The most important spatial criterion is the IACC. The shape of the hall should be adjusted to minimize the IACC and maximize the subjective preference at each seat.

A. *Fundamental Schema*

To determine the fundamental space shape, the IACC was calculated by using the image method for a concert hall with width W, length L and height H (Fig. 7.4.1, from Miyamoto and Ando, unpublished). The floor area was fixed at a certain value, and the stage area was fixed at 1/3 of the seating area. As discussed by Miyamoto and Ando (unpublished), the seat floor was inclined by 12° in order to minimize the attenuation of direct sound. For simplification, the sound source was placed on the centre of the stage (1.5 m above the floor, 20% of the distance from the front of the stage). The absorption coefficient of the seating area was assumed to be 0.65. Sound simulation was performed at 80–100 seat positions with the receiving point at the height of a listener's ear, 1.1 m. The distribution of the IACC value calculated (using music motif B) is shown in Fig. 7.4.2. It is generally known that the IACC is increased near the sound source. Narrowing the hall width W decreased the IACC and increased the seating area in which IACC < 0.5. This relates to the importance of sound reflected from the side walls.

B. *Ceiling*

As shown in Figs 7.4.3 and 7.4.4, the effects of varying the ceiling surface angle were examined. Figure 7.4.3 shows the ceiling angle Ψ for the auditorium at Kobe University (which was changed). Figure 7.4.4 is for Boston Symphony Hall, with the heights $H = 18$ m (a) and $H = 9$ m (b). The IACC values at specified seats were calculated for the sound source located at the centre of the stage. Figure 7.4.5 shows the scale value of subjective preference S_4 due to the IACC value calculated using Eq. (5.1.13). When the ceiling angle $\Psi = 30°$, the sound field of the auditorium at Kobe University is improved to a level similar to that in the Boston Symphony Hall.

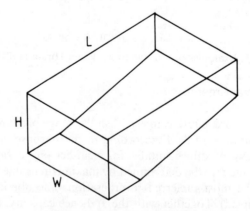

Fig. 7.4.1. Geometry of a concert hall.

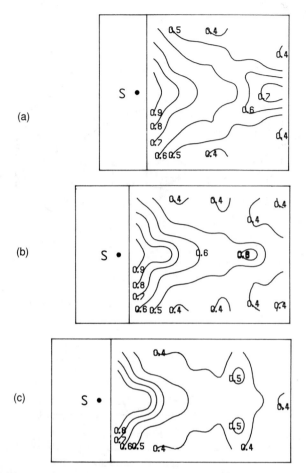

Fig. 7.4.2. Calculated values of IACC: (a) $L/W = 1.2$, $H/W = 0.6$; (b) $L/W = 1.6$, $H/W = 0.6$; (c) $L/W = 2.0$, $H/W = 0.6$.

C. Side Walls near the Sound Source

Sound fields were next simulated when changing the shape of the side walls on the stage and near the stage. Calculated IACC values are shown in Fig. 7.4.6 for an initial scheme and for an alternative scheme (Ando et al., 1992). As shown in Fig. 7.4.6b, the alternative scheme obviously results in decreasing the seating area with IACC values greater than 0.5. The resulting total amplitude of reflections, however, as well as the listening level and the initial time delay between the direct sound and the first reflection were not changed. Thus, the total scale values of subjective preference according to the four factors with Eq. (5.1.12) depend mainly on the IACC. Typical results are shown in Fig. 7.4.7 for fast music (motif B: Sinfornietta, Opus 48; III movement composed by Malcolm Arnold with effective duration of ACF of $\tau_e = 43$ ms). Plate III indicates each physical factor at each seat, and Plate IV indicates the subjective preference corresponding to each orthogonal factor and the total preference at each seat. Sound fields at most of the seats are from -0.5 to -0.8. Note that the scale value is not zero at any seat, since the IACC value may not be 'zero'.

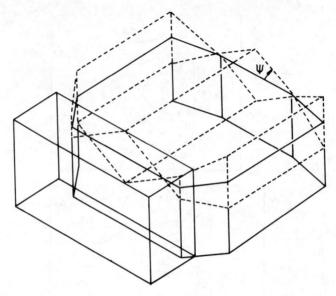

Fig. 7.4.3. Auditorium at Kobe University (ceiling angle Ψ varied in the simulation).

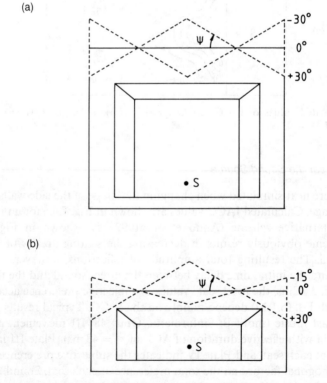

Fig. 7.4.4. Boston Symphony Hall with ceiling angle Ψ varied in the simulation: (a) $H = 18\,\mathrm{m}$ (original); (b) $H = 9\,\mathrm{m}$ (varied).

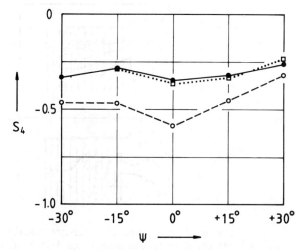

Fig. 7.4.5. Scale values S_4 due to IACC (music motif B): broken line, auditorium at Kobe University; solid line, Boston Symphony Hall, $H = 18\,\text{m}$; dotted line, Boston Symphony Hall, $H = 9\,\text{m}$.

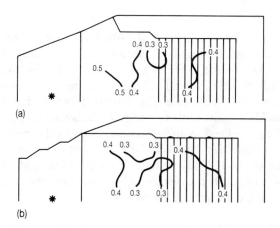

Fig. 7.4.6. Contour lines of equal IACC calculated by Eq. (5.1.7) with correlation values listed in Ando (1985) with music motif B: (a) initial scheme; (b) alternative scheme.

D. *Schroeder's Diffuser*

In the final design of the concert hall, Schroeder's diffusers were added to the central part of the ceiling in order to prevent strong reflections from the median plane (Fig. 7.4.8). The diffuser, with a designed frequency range of 250–1750 Hz (Schroeder, 1979), is shown in Fig. 7.4.9. After construction of the hall, the IACC was measured for several seat positions (Fig. 7.4.8). In the design stage, the IACC could not be calculated with the Schroeder diffusers (Fig. 7.4.6), but it is interesting that at the centre parts of the hall, where the diffusers close to the stage may be effective, the measured IACCs are much smaller than those calculated (Fig. 7.4.6(b)).

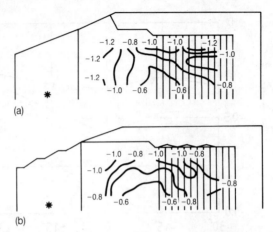

Fig. 7.4.7. Contour lines of equal total scale value of preference calculated by Eqs (5.1.12) through (5.1.17): (a) initial scheme: (b) alternative scheme. From Ando et al. (1992).

7.4.3. New Design for Reflectors

A. *Triangular Reflectors*

In another hall designed by Nakajima et al. (1992), triangular reflectors were installed above and near the stage. Triangular reflectors with an angle of about 120° show effective reflections over a wide range of frequencies. When such reflectors are placed near the ceiling on both sides, lateral reflections in the low-frequency range may be obtained to decrease the IACC (see Fig. 5.6.9). It seems that the canopy in the Tanglewood Music Shed satisfies this condition (Beranek, 1962). Note that for high frequencies, the effective reflection angle becomes close to the median plan of listeners (Fig. 5.6.9). This condition helps to avoid the image shift of the sound source and keeps the maximum value of the ICF at the time of origin. Measured IACCs are shown in Fig. 7.4.10. When the reflectors are installed above and near the stage, the IACC values of seats close to the stage are greatly decreased.

B. *Fractal Structures*

For a wider range of reflected frequencies, the fractal structures shown in Fig. 7.4.11 are proposed for triangular reflectors (a) as well as for Schroeder's reflectors (b). To control the proper reflection angle for each frequency band and for each listener according to Fig. 5.6.9, a fractal structure (Mandelbrot, 1983; Schroeder, 1991) like that shown in Fig. 7.4.11(c) may be useful. Calculated results are shown in Fig. 7.4.12 (Ando and Sakamoto, 1988): three directions (for the low-, middle-, and high-frequency ranges) may be realized.

7.4.4. Design of an Electroacoustic System in an Enclosure with Multiple Channels of Loudspeakers

An electroacoustic system with multiple channels of loudspeakers may be designed in a manner similar to that used in designing a concert hall with music sources as well as

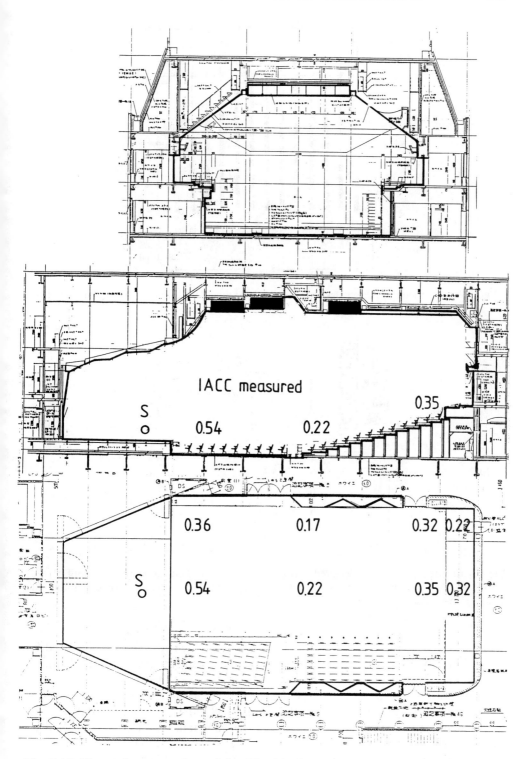

Fig. 7.4.8. Plan and cross sections of the Higashinada-Kumin Centre, Kobe, with measured IACCs.

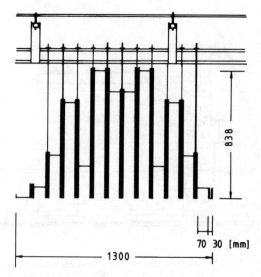

Fig. 7.4.9. Schroeder's diffuser designed for the Higashinada-Kumin Centre, Kobe ($N = 13$).

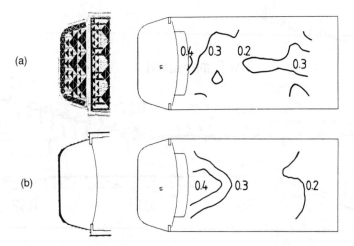

Fig. 7.4.10. Measured and calculated IACCs in the Fujita Hall 2000 (music motif B). (a) Measured IACCs with the canopy. (b) IACCs calculated by the image method (without the canopy). From Nakajima et al. (1992).

speech sources on the stage. The impulse response at each seat can be calculated by taking into consideration the directivities of loudspeakers, the properties of the delay machines and reverberators, and the reflection properties of the walls. Thus the global subjective preference according to the four acoustic factors may be obtained. For a public address system, for example, the range of effective duration of ACF (τ_e) for speech signals may be 10–30 ms. For music reproduction, the range of τ_e should be decided according to the purpose of the acoustic space. The range of τ_e for music is usually much greater than that for speech, so two loudspeaker systems are recommended for maximizing subjective preference. If the subjective preference is maximized

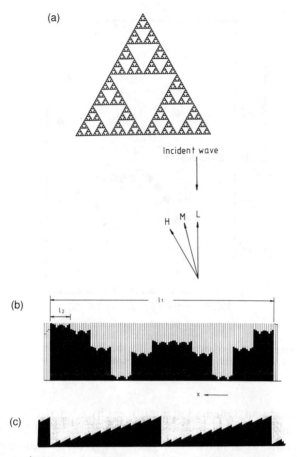

Fig. 7.4.11. Fractal geometries of reflectors for wide frequency ranges. (a) Fractal reflectors suggested for the wide frequency range. (b) Schroeder's diffusers with fractal structure ($N = 11 \times 11$). (c) Fractal structure for controlling directions of reflection for different frequency ranges ($N = 121$).

for speech, speech clarity may also be increased by adjusting the temporal and spatial factors to minimize IACC (Section 5.6.2).

7.4.5. Seat Selection for Individual Listener

As mentioned in Section 5.5, large individual differences were observed particularly for LL, Δt_1, and T_{sub}. The large individual differences in the most preferred listening level are at least partly related to the individual hearing level, and the preferred initial time delay and the preferred reverberation time are associated with 'liveness'. Taste, therefore, depends greatly on the individual, and a seat selection system is used to fit the preference of each individual to the sound field described by the four acoustic factors. Preference tests can be performed in a manner similar to that used for testing eyesight for the selection of spectacles. A suitable method would be to use either the

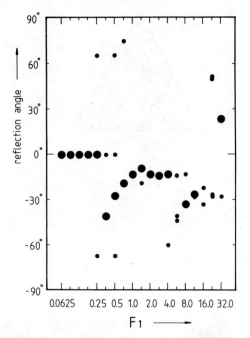

Fig. 7.4.12. Calculated reflection characteristics of the fractal structure (Fig. 7.4.11c) as a function of the normalized frequency. (Ando and Sakamoto, 1988.)

paired-comparison test and a method of adjustment by a listener or a test person, or a combined method that obtains the most preferred combination of LL, Δt_1 and T_{sub}. To examine the preference as a function of IACC, only the paired-comparison test is applicable, since there is no peak of preference against the IACC. The first simulation system for this purpose was designed for the Kirishima International Music Hall, Kagoshima Prefecture in Japan (Plate V), and opened in 1994.

EXERCISES

Chapter 2

1. Show that the Fourier transform of a convolution of $x_1(t)$ and $x_2(t)$ is equal to $X_1(f)X_2(f)$.
2. Show that the Fourier transform of a correlation of $x_1(t)$ and $x_2(t)$ is equal to $X_1^*(f)X_2(f)$.
3. Prove the transform pair given by Eq. (2.1.12).
4. Prove the transform pair given by Eq. (2.1.14).
5. Prove the transform pair given by Eq. (2.1.26).
6. Confirm DFT pairs given in Fig. 2.2.1.
7. Show that $X(k) = X^*(N - k)$ if $\{x(n)\}$ is real, where $x(n)$ and $X(k)$ are defined by Eq. (2.2.1) and Eq. (2.2.2).
8. Prove Eq. (2.3.3) by use of the method of the calculus of residues.
9. Show that the phase of $H_2(f)$ given by Eq. (2.3.18) is equal to $+\pi$ and $-\pi$ at $f = -\infty$ and ∞, respectively.
10. Prove Eq. (2.3.33) assuming that $f_m < f_c$.
11. Derive Eq. (2.3.35) using the definite integral formula;

$$\int_{-\infty}^{+\infty} \frac{\sin(ax)}{x(x - b)} \, dx = \pi\{\cos(ab) - 1\}/b.$$

Chapter 3

1. Show examples that illustrate the changes of the transfer function as the number of reflection sounds increases. Where are the poles and zeros in your examples of transfer functions?
2. What are the conditions that the mass law holds well for the transmission loss of a single wall?
3. Consider the differences between impulse responses in a dispersive medium and in a nondispersive medium.
4. Describe the assumptions for Sabine's and Eyring's reverberation formulae, respectively.

303

5. Consider the effects of directivity of a sound source on the reverberation decay in a 2-dimensional or almost-2-dimensional diffuse field.
6. Consider the relationship between the ray theory and the wave theory in a closed space. Where do you see the eigen frequencies in the space based on the ray theory?
7. The sound field has a deterministic nature following a linear wave equation. What is the basic idea for the statistical analysis of sound fields?
8. Consider the poles and zeros of a transfer function that has complex modal functions.
9. Consider the effects of truncation and sampling of an impulse response data on the pole-zero distribution. Where do you see the poles of a transfer function in your measured transfer function data?

Chapter 4

1. Discuss the locus of the particle velocity in Fig. 4.2.2.
2. Discuss a method to reverse the direction of the energy flow in Fig. 4.2.4.
3. Discuss a reason(s) for the clockwise rotation of sound energy in Fig. 4.3.2.
4. Use Fig. 4.3.3 to explain conditions for one of two monopoles, that are vibrating in-phase to each other, to absorb sound energy.
5. Prove Eq. (4.3.10).
6. Derive Eqs. (4.3.12) and (4.3.13).
7. Explain reasons why convex and concave domes have wider directivities than a flat piston with the same radius.
8. Explain why #2 in Fig. 4.3.20 has only positive regions and #3 in the same figure has positive as well as negative regions.
9. Obtain Eq. (4.4.9) by use of Eqs. (4.1.18) and (4.1.21).
10. Discuss the similarities and differences between the envelope intensity and ISIS.

Chapter 5

1. Consider an acoustic measurement system for sound field in rooms and a "diagnostic system" calculating subjective attributes at each seat
2. Investigate temporal and spatial factors associated with the left and right hemispheres for visual design of the environment.

Chapter 6

1. What is the difference between frequency characteristics of transfer functions and power responses?
2. Describe the relationship between the sound power output of a source and space-time-averaged mean squared sound pressure in a closed space.
3. Consider relationship between the subjective diffuseness and statistical properties of binaural signals.
4. Show a concrete example that demonstrates instability or noncausality of an inverse filter for a nonminimum phase transfer function.
5. How does one extract the minimum phase components of an impulse response?
6. Consider the cepstrum windowing effects on transfer function compared with time windowing effects of an impulse response on the transfer function.
7. Consider the distribution of poles and zeros of the transfer function of a cascaded-linear system.
8. Describe advantages of pole-zero modelling of a TF compared with all-zero modelling.
9. Consider a generating function for a sequence as a method of transfer function modelling.
10. Consider the energy input–output relationship between a primary source and a secondary source used for active power minimization in a closed space.
11. How does one detect the changes in the transfer function and estimate a new transfer function?

Chapter 7

1. Consider all the orthogonal acoustic factors that influence in the judgement of spatial impression of sound fields in a room.
2. Show an example of a plan and cross sections of proposed concert hall maximizing the global scale value of subjective preference at each seat for chamber music like that of Mozart in which the effective duration of the autocorrelation function is assumed to be centred on, say, about 65 ms.
3. Design an electroacoustic system for speech signals incorporating temporal and spatial acoustic factors in an auditorium.
4. How does one measure speech intelligibility under reverberant conditions?
5. Consider binaural effects on speech intelligibility.
6. Why can we detect sound source location under reverberant conditions? What conditions are necessary for a sound source to be able to detect its location in a reverberant space?

APPENDICES

Appendix 1: Derivation of $P_{s,\text{zero-cloud}}$ in Equation 3.6.25

The power spectrum of the group delay due to the zero-cloud singularities is given by

$$P_{s,\text{zero-cloud}}(x) = \frac{n_p}{4\pi} \int_0^\infty \exp(-2x\beta'/n_p) \left(\frac{1}{1 + 4(\beta' + \beta_0)^2} + \frac{1}{1 + 4(\beta' - \beta_0)^2} \right) d\beta'$$

$$= \frac{n_p}{4\pi} (I_1 + I_2) \tag{A1.1}$$

where

$$I_1 = \int_0^\infty \exp(-2x\beta'/n_p) \frac{1}{1 + 4(\beta' + \beta_0)^2} \, d\beta' \tag{A1.2a}$$

and

$$I_2 = \int_0^\infty \exp(-2x\beta'/n_p) \frac{1}{1 + 4(\beta' - \beta_0)^2} \, d\beta' \tag{A1.2b}$$

The integration above is rewritten by integration by parts,

$$I_1 = \int_0^\infty \exp(-2x\beta'/n_p) f_1(\beta') \, d\beta'$$

$$= \left[-\frac{n_p}{2x} \exp(-2x\beta'/n_p) f_1(\beta') \right]_0^\infty + \frac{n_p}{2x} \int_0^\infty \exp(-2x\beta'/n_p) f_1'(\beta') \, d\beta'$$

$$= \frac{n_p}{2x} \left(\frac{1}{1 + 4\beta_0^2} \right) - \left(\frac{n_p}{2x} \right)^2 [\exp(-2x\beta'/n_p) f_1'(\beta')]_0^\infty$$

$$+ \left(\frac{n_p}{2x} \right)^2 \int_0^\infty \exp(-2x\beta'/n_p) f_1''(\beta') \, d\beta'$$

$$\sim \frac{1}{8\beta_0^2} \left(\frac{n_p}{x} \right) \qquad \left(\frac{n_p}{2x} \to \text{small} \right) \tag{A1.3}$$

306

where

$$f_1(\beta') = \frac{1}{1 + 4(\beta' + \beta_0)^2} \tag{A1.4}$$

Consequently, we have

$$P_{s,\text{zero-cloud}}(x) \approx \frac{1}{16\pi\delta_0}\left(\frac{1}{\delta_0 x}\right) \tag{A1.5}$$

Appendix 2:
Zero-Counting in
a Frequency Band

Consider a closed area in the z-plane, as shown by the hatched region in Fig. A2.1, which is enclosed by two circles with radii $r_1, r_2 (r_1 < r_2)$ and two radial axes with directions ω_1 and ω_2. If $H(z)$ is analytic on the enclosure and inside the area, then the contour integral of $H'(z)/H(z)$ around the enclosed region C will be related to the number of zeros and poles of $H(z)$ located in the area, i.e.

$$\frac{1}{2\pi j}\int_c \frac{H'(z)}{H(z)}\,\mathrm{d}z = N_z - N_p \tag{A2.1}$$

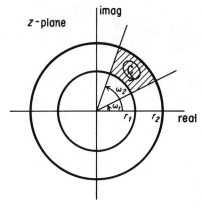

Fig. A2.1. z-plane for zero-counting. From Tohyama et al. (1991).

where N_z is the number of zeros and N_p denotes the number of poles. Suppose that

$$H(z) = \sum_{k=0}^{N-1} a_k z^{-k} \qquad \text{(A2.2)}$$

then $H(z)$ has N^{-1} order poles at the origin and N^{-1} zeros over the entire z-plane, so that $N_p = 0$ for $r_1 > 0$ in Eq. (A2.1). Therefore, N_z contained inside C can be obtained by carrying out the contour integration of the left-hand side in Eq. (A2.1). The contour integral consists of the integrals over two arcs and two radial axes.

The integral over an arc with radius r can effectively be calculated numerically by using the FFT algorithm in the following way.

$$\frac{1}{2\pi j} \int_{\text{arc}} \frac{H'(z)}{H(z)} \, dz = \frac{1}{2\pi} \int_{\omega_1}^{\omega_2} \frac{\sum_{k=0}^{N-1} -k a_k r^{-k} e^{-j\omega k}}{\sum_{k=0}^{N-1} a_k r^{-k} e^{-j\omega k}} \, d\omega$$

$$\approx \frac{1}{M} \sum_{k=k_1}^{k_2} \frac{DFT_{(M)}(c_k)}{DFT_{(M)}(b_k)} \qquad \text{(A2.3)}$$

where

$$b_k = \begin{cases} a_k r^{-k} & (0 \leqslant k \leqslant N - 1) \\ 0 & (N \leqslant k \leqslant M - 1) \end{cases}$$

$$c_k = \begin{cases} -k a_k r^{-k} & (0 \leqslant k \leqslant N - 1) \\ 0 & (N \leqslant k \leqslant M - 1) \end{cases}$$

$$k_i = \left(\frac{\omega_i}{2\pi} M \right) \qquad (i = 1, 2)$$

$DFT_{(M)}$ = M-point discrete Fourier transform.

$M > N$ would be preferable in order to calculate Eq. (A2.2). If $M = 2^L$ (L integer), the FFT algorithm is useful. On the other hand, for the integral over a radial direction, there may be no appropriate effective calculation algorithm and we carry out the next numerical integration.

$$\frac{1}{2\pi j} \int_{\text{radial}} \frac{H'(z)}{H(z)} \, dz \approx \frac{1}{2\pi j} \sum_{r=r_1}^{r_2} \frac{\sum_{k=0}^{N-1} -k a_k r^{-k-1} e^{-j\omega k}}{\sum_{k=0}^{N-1} a_k r^{-k} e^{-j\omega k}} \, \Delta r \qquad \text{(A2.4)}$$

For example, in the case of zero-counting in 500 Hz (1/1 octave band width), an impulse response record has been acquired by sampling at frequency of 1414 Hz. After removing the pure delay part observed at the top of the impulse response data, 4096 ($= N$) points of data were chosen starting from the head of the resulting impulse response record, and 0-value samples were padded to get a data length of $M = 2^{16}$ for further FFT processing in Eq. (A2.2). In the calculation of Eq. (A2.3), we used Δr larger than 0.0001, which is variable depending on the magnitude of r. The total number M of data including zero padding is fixed for all the calculations. The sampling frequency and the impulse response record length are changed by the frequency band of interest and the reverberation time.

Appendix 3:
Group Delay Analysis

A. Group Delay Calculation

The measured transfer functions, of length originally 16 384 samples, are processed using the procedure shown in Fig. A3.1. First, the pure delay part observed at the top of the TF is eliminated, and then the TF is truncated so that the total data length becomes a particular length that is determined from the reverberation time (T_R) of the frequency bond.

The data lengths of the three frequency bands are listed in Table A3.1 together with the T_R of the room and the sampling frequency used to acquire the TFs. The data length for each frequency band is longer than the reverberation time T_R. Then zero-padding of the truncated TF is performed to get a record of length 2^{14} for the following FFT operation. The number of FFT frequency points is 2^{14} for all the frequency band TFs.

After the FFT, the continuous phase function, $\phi(n)$, is formed using a phase-unwrapping technique, followed by removing the linear phase component. The group delay function is defined here as the first derivative of the resulting phase function.

Let $\phi_0(n)$ be a discretely sampled expression of a continuous phase function, then the group delay function $\tau(n)$ that we use is defined as follows:

$$\tau(n) = \frac{\phi_0(n) - \phi_0(n-2)}{2\Delta\omega} \quad \text{(s)} \tag{A3.1}$$

where $\Delta\omega$ is the angular frequency of sampling $\phi_0(n)$. There is an alternative way of obtaining the first derivative in which the difference is taken between two adjacent samples and divided by $\Delta\omega$. The obtained group delay function has much noise at the high frequency region, presumably because of the aliasing introduced by sampling. This high frequency noise can be suppressed by a two-sample moving-average operation. That is why we use $\tau(n)$ as defined above.

Although there are 2^{14} samples of $\phi_0(n)$ and $\tau(n)$, the effective amount of data we could process is 2^{12} samples, because only the half of the positive frequency band data is useful for 1/1 octave signals.

B. Power of Group Delay

To analyse the power of $\tau(n)$, the 4096 points ($n = 0, 1, \ldots, 4095$) are divided into W sections of length L, i.e. $WL = 4096$. Then we calculate the power of the ith section, $P_s^{(i)}(x)$,

$$P_s^{(i)}(x) = \left| \left(\sum_{n=0}^{L-1} \tau_i(n) e^{j(2\pi/L)nx} \right) \right|^2 \tag{A3.2}$$

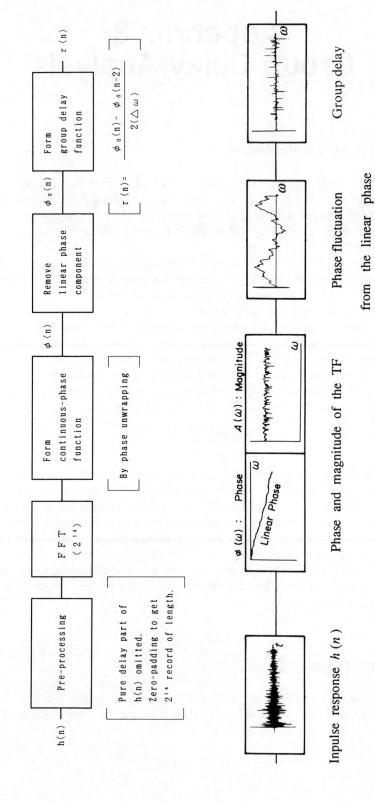

Fig. A3.1. Procedure for obtaining phase and group delay functions. From Tohyama et al. (1994).

Table A.3.1. Data table for the processing. After Tohyama et al. (1994)

1/1 Octave band (Hz)	TF data length	Sampling frequency (1/s)	T_R (s) (no. of samples)
			2.05
250	2048	707	(1449)
			1.81
500	4096	1414	(2559)
			1.65
1000	8192	2828	(4666)

Then we take the ensemble average of $P_s^{(i)}(x)$ over W sections, i.e.

$$P_s(x) = \langle P_s^{(i)}(x) \rangle_i \qquad (A3.3)$$

The power data in Fig. 3.6.12 are for $L = 256$ samples and $W = 16$. $L = 256$ corresponds to the real frequency 21.8 Hz for the 500-Hz 1/1 frequency band.

C. Δ-Statistics of the phase function $\phi(n)$

The discrete version of Eq. (3.6.32) is

$$\Delta = \min \frac{1}{L} \sum_{n=0}^{L-1} \langle \phi_0(n) + An - B \rangle^2 \qquad (A3.4)$$

where L is the number of data points in a particular interval. Suppose we use the theoretical value for A, then B that minimizes the right-hand side of Eq. (A3.4) is

$$B = \frac{1}{L} \sum_{n=0}^{L-1} \langle \phi_0(n) + An \rangle \qquad (A3.5)$$

For the calculation, the 4096 $\phi_0(n)$ data points are divided into W sections of length L. Expressing $\phi_0(n)$ in the ith section by $\phi_{0i}(n)$, and B by B_i, then the Δ-statistic of the ith section Δ_i is obtained by

$$\Delta_i = \frac{1}{L} \sum_{n=0}^{L-1} \langle \phi_{0i}(n) + An \rangle^2 - B_i^2 \qquad (A3.6)$$

After this we take the ensemble average of Δ_i over W sections ($N = 16$, $L = 256$).

Appendix 4: Minimum Phase Components Decomposition and Inverse Filtering Procedure

In Fig. A4.1, the left branch in the figure shows the inverse filter creation process and the right branch shows the source waveform recovery process by inverse filtering.

An impulse response signal, measured between particular source and receiver locations, is processed to give an inverse filter. Another response signal, measured under the same conditions, is inversely filtered to recover the source waveform. The former response will be referred to as the reference response and the latter as the test response. Both the reference and test responses have length (N_L) of 4096 data records based on a 1.8 s reverberation time (500 Hz 1/1-octave frequency band). Our sampling frequency is 1414 (1/s).

The reference response is first exponentially weighted, and then the magnitude $A(\omega)$ and phase function $\phi_w(\omega)$ are obtained by Fourier transformation. $\phi_w(\omega)$ is changed (unwrapped) to obtain a continuous function of frequency $\phi(\omega)$. This is necessary because a phase function calculated as an argument of the complex Fourier transform is not a continuous frequency function. The magnitude function $A(\omega)$ is unchanged. Figure A4.2 shows a portion of the response in a very narrow frequency interval in the 1/1-octave-frequency band.

The linear phase component in the unwrapped phase function is eliminated. The tangent of the linear phase function is calculated from the difference between the unwrapped phase values at both ends of the frequency band, divided by the frequency interval. The resultant phase function has zero values at both ends of the frequency band.

The group delay function is calculated by taking the difference between adjacent phase sample values. Let $\phi_0(n)$ be the phase with the linear phase eliminated and $\tau(n)$ be the group delay function:

$$\tau(n) = \frac{\phi_0(n) - \phi_0(n-1)}{\Delta\omega} \tag{A4.1}$$

In this equation, $\Delta\omega$ is the frequency resolution of $\phi_0(n)$, and in our experiment it is $2\pi(1414/16384)$ (rad/s). The number of excess data over 4096 in the response record is due to the zero padding.

In the next step, the sign of the negative values of the group delay function obtained is reversed to give an absolute magnitude of the group delay function. The absolute value group delay function is then integrated to again create a phase function by using

$$\phi(n) = \phi(n-1) + |\tau(n)|, \qquad \phi(0) = 0 \tag{A4.2}$$

The resulting phase function $\phi(n)$ has a linear phase component, so as a final step we eliminate it in the same way as mentioned in the description of linear phase

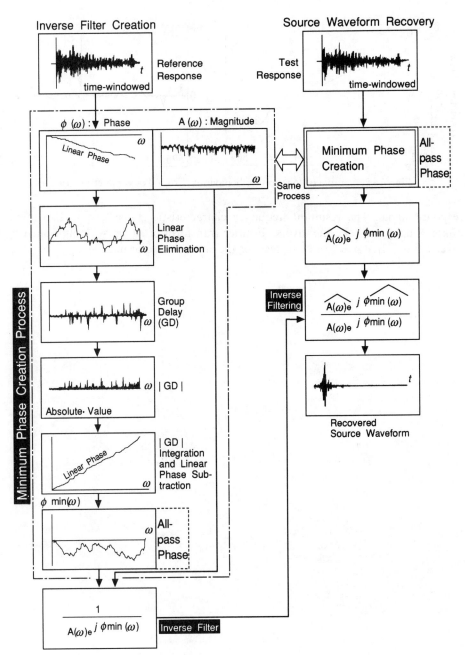

Fig. A4.1. Inverse filtering procedure. From Tohyama et al. (1992b).

elimination of the original unwrapped phase function. The resultant phase function $\phi_{\min}(\omega)$ is considered to be the minimum phase of the reference response. We can get an inverse filter whose frequency characteristic is $(1/A(\omega))\mathrm{e}^{-\mathrm{j}\phi_{\min}(\omega)}$.

Source waveform recovery is performed by the procedure shown in the right branch of Fig. A4.1. The test response is processed in exactly the same way as for the reference

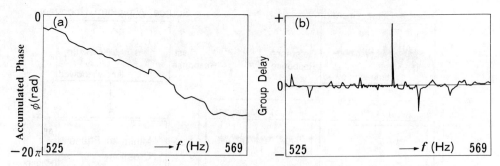

Fig. A4.2. Sample of local phase behaviour. From Tohyama et al. (1992a).

response signal. The resultant frequency characteristic $\widehat{A(\omega)}e^{j\phi_{min}(\omega)}$ is also minimum phase. Finally, we take the inverse Fourier transform of $\widehat{A(\omega)}e^{j\phi_{min}(\omega)}$ multiplied by the inverse filter $(1/A(\omega))e^{-j\phi_{min}(\omega)}$, to recover the source waveform.

Appendix 5: Cepstrum Dereverberation Procedure

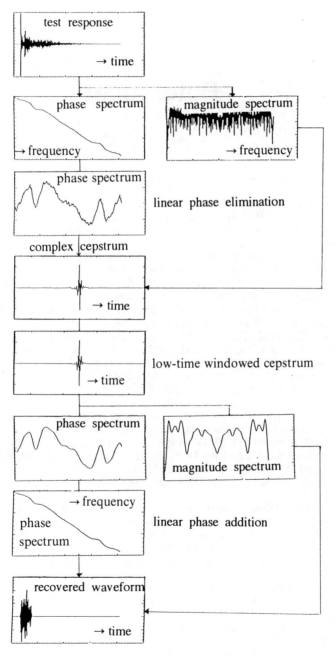

Fig. A5.1.

References

Alrutz, H. and Schroeder, M. R. (1983) *Proc. 11th Int. Conf. Acoustics,* Paris, 235–238.
Ando, Y. (1977) *J. Acoust. Soc. Am.* **62**, 1436–1441.
Ando, Y. (1983) *J. Acoust. Soc. Am.* **74**, 873–887.
Ando, Y. (1985) *Concert Hall Acoustics.* Springer-Verlag, Berlin.
Ando, Y. (1986) *Proc. 12th Int. Congr. Acoustics, Toronto,* p. E4-14.
Ando, Y. (1992) *Acustica* **76**, 292–296.
Ando, Y. and Alrutz, H. (1982) *J. Acoust. Soc. Am.* **71**, 616–618.
Ando, Y. and Kurihara, Y. (1986) *J. Acoust. Soc. Am.* **80**, 833–836.
Ando, Y. and Morioka, K. (1981) *J. Acoust. Soc. Jpn* **37**, 613–618 (in Japanese).
Ando, Y. and Sakamoto, M. (1988) *J. Acoust. Soc. Am.* **84**, 1734–1740.
Ando, Y., Shidara, S. and Maekawa, Z. (1974) *Proc. Int. Congr. Acoustics, London,* p. 611.
Ando, Y., Okura, M. and Yuasa, K. (1982) *Acustica* **50**, 134–141.
Ando, Y., Kang, S. H. and Morita, K. (1987a) *J. Acoust. Soc. Jpn (E)* **8**, 197–206.
Ando, Y., Kang, S. H. and Nagamatsu, H. (1987b) *J. Acoust. Soc. Jpn (E)* **8**, 183–190.
Ando, Y., Yamamoto, K. et al. (1991) *Acoust. Letters* **15**, 57–64.
Ando, Y., Yamamoto, M. et al. (1992) *Proc. 14th Int. Congr. Acoustics,* p. F4-3.
Aoki, S. and Houtgast, T. (1992) *Hearing Res.* **59**, 25–30.
Aoki, S., Miyata, H and Sugiyama, K. (1990) *J. Audio Eng. Soc.* **38**, 433–439.
Baxter, S. M. and Morfey, C. L. (1986) *Angular Distribution Analyis in Acoustics*, Lecture Notes in Engineering, no. 17. Springer-Verlag, Berlin.
Barron, M. (1973) *J. Sound Vib.* **27**, 183–196.
Bartel, T. W., Yaniv, S. L. and Flynn, D. R. (1983) *J. Acoust. Soc. Am.* **80**, 828–832.
Batchelder, L. (1964) *J. Acoust. Soc. Am.* **36**, 551–555.
Bendat, J. S. and Piersol, A. G. (1971) *Random Data: Analysis and Measurement Procedures.* Wiley, New York.
Beranek, L. L. (1962) *Music, Acoustics, and Architecture.* Wiley, New York.
Blauert, J. (1983) *Spatial Hearing.* MIT Press, Boston, MA.
Borish, J. (1985) *J. Audio Eng. Soc.* **33**, 888–891.
Bolt, R. H. (1947) *J. Acoust. Soc. Am.* **19**, 79–90.
Botte, M. C., Bujas, Z. and Chocholle, R. (1975) *J. Acoust. Soc. Am.* **58**, 208–213.
Buchwald, J. S. and Huang, C. M. (1975) *Science* **189**, 382–384.
Burkhard, B. and Sachs, R. (1978) *J. Acoust. Soc. Am.* **58**, 214–222.
Choi, H. I. and Williams, W. J. (1989) *IEEE Trans. ASSPV* 862–871.
Chu, W. T. (1982) *J. Acoust. Soc. Am.* **72**, 196–199.
Chung, J. Y. (1977) Research Publication, General Motors Research Laboratory, BMR-2617. Warren, MI.
Claasen, T. A. C. M. and Mecklenbrauker, W. F. G. (1980) *Philips J. Res.* **35**, 217–250, 276–300 and 372–389.

Cohen, L. (1966) *J. Math. Phys.* **7**, 781–786.

Cohen, L. (1989) *Proc. IEEE* **77**, 941–981.

Combes, J. M., Grossmann, A. and Tchamitchian, P. H. (1989) *Wavelets, Time-Frequency Methods and Phase Space.* Springer-Verlag, Berlin.

Cook, R. K., Waterhouse, R. V., Benrendt, R. D. et al. (1955) *J. Acoust. Soc. Am.* **27**, 1072–1077.

Cooley, J. W. and Tukey, J. W. (1965) *Mathematical Computations* **19**, 297–301.

Cremer, L. (1949) *Die Wissenschaftlichen Grundlagen der Raumakustik.* Hirzel Verlag, Stuttgart.

Damaske, P. (1967) *Acustica* **19**, 199–213.

Damaske, P. (1971) *J. Acoust. Soc. Am.* **50**, 1109–1115.

Damaske, P. and Ando, Y. (1972) *Acustica* **27**, 232–238.

Davy, J. (1981) *J. Sound Vib.* **77**, 455–479.

Davy, J. (1990) *Inter-Noise '90*, Gothenburg, 159–164.

Dubrovskii, N. A. and Chernyak, R. I. (1969) *Soviet Physics — Acoustics* **14**, 326–332.

Dyson, F. and Mehta, M. (1963) *J. Math. Phys.* **4**, 701–712.

Ebeling, K. (1985) *Physical Acoustics, XVII*, pp. 233–310. Academic Press, London.

Eyring, C. F. (1930) *J. Acoust. Soc. Am.* **1**, 217–241.

Fahy, F. J. (1977) *Noise Control Eng.* **9**, 155–162.

Fahy, F. J. (1989) *Sound Intensity.* Elsevier Applied Science, London.

Fahy, F. J. (1992) *Proc. 2nd Int. Congr. Recent Developments in Air- and Structure-borne Sound and Vibration,* Auburn, pp. 611–618.

Fujii, K. and Ohga, J. (1992) *Trans. IEICE* **E75-A**, 1509–1515.

Gabriel, K. and Colburn, H. (1981) *J. Acoust. Soc. Am.* **69**, 1394–1401.

Gold, B. and Rader, C. M. (1969) *Digital Processing of Signals.* McGraw-Hill, New York.

Gotoh, T., Kimura, Y., Yamada, A. and Watanabe, K. (1984) *J. Acoust. Soc. Jpn (E)* **5**, 85–94.

Gradshteyn, I. S. and Ryzhik, I. M. (1965) *Tables of Integrals, Series, and Products.* Academic Press.

Haas, H. (1951) *Acustica* **1**, 49–58.

Haas, H. (1972) *J. Audio Eng. Soc.* **20**, 145–159.

Haneda, Y., Makino, S. and Kaneda, Y. (1994) *IEEE Trans. Speech Audio, Processing* **2**, 320–328.

Harris, F. J. (1978) *Proc. IEEE* **66**(1), 51–83.

Haykin, S. (1991) *Adaptive Filter Theory*, 2nd edn. Prentice Hall, Englewood Cliffs, N.J.

Hirata, Y. (1979) *Acustica* **43**, 247–252.

Hirata, Y. (1981) *J. Audio Eng. Soc.* **29**, 333–337.

Hirata, Y. (1982a) *J. Sound Vib.* **82**, 593–595.

Hirata, Y. (1982b) *J. Sound Vib.* **84**, 509–517.

Hirobayashi, S. (1993) Master Thesis, Kogakuin Univ.

Hirobayashi, S., Tohyama, M., Koike, T., and Lyon, R. H. (1994) *J. Acoust. Soc. Am.* **95** (No. 5, Pt. 2) 2901 (3aSAa10).

Houtgast, T. and Steeneken, H. (1985) *J. Acoust. Soc. Am.* **77**, 1069–1077.

Houtgast, T., Steeneken, H. J. M. and Plomp, R. (1980) *Acustica* **46**, 60–72.

Iizuka, M., Turumaki, H., Aoki, M. and Suzuki, M. (1991) *Tech. Rep. IEICE Jpn* **EA91-75** (in Japanese).

Imai, A., Inazumi, A. and Konishi, M. (1992) *Tech. Rep . IEICE Jpn* **EA92-38** (in Japanese).

Ino, T., Okamoto, K., Yoshida, A. and Inoue, J. (1985) *Audiology Jpn* **28**, 251–256.

Itow, T. (1957) *Onkyo Kogaku Genron* (in Japanese). Corona, Tokyo.

Janse, C. P. and Kaizer, J. M. (1983) *J. Audio Eng. Soc.* **37**, 198–223.

Jasper, H. H. (1958) *Electroenceph. Clin. Neurophysiol.* **10**, 371–375

Jewett, D. L. (1970) *Electroenceph. Clin. Neurophysiol.* **28**, 609–618.

Kaneda, Y. and Ohga, J. (1986) *Proc. IEEE* **ASSP34**, 1391–1400.

Katsuki, Y., Sumi, T., Uchiyama, H. and Watanabe, T. (1958) *J. Neurophysiol.* **21**, 569–588.

Kawakami, F. and Yamaguchi, K. (1986) *J. Acoust. Soc. Am.* **80**, 543–554.

Kawaura, J., Suzuki, H. and Ono, T. (1987) *Nikkei Electronics* **423**, 163–177.

Keet, M. V. (1968) *Proc. 6th Int. Congr. Acoustics,* p. E-2-4.

Kido, K. (1988) *Acoustics.* Corona, Tokyo.

Kim, J. T. and Lyon, R. H. (1988) *Proc. Noise-Con '88*, 493–498.

Kim, J. T. and Lyon, R. H. (1992) *J. Mech. Syst. Signal Proc.* **6**, 1–15.

Kinsler, L. E. and Frey, A. R. (1962) *Fundamentals of Acoustics.* Wiley, New York.

Koizumi, N. and Lyon, R. H. (1989) *J. Acoust. Soc. Am.* **86** (suppl. 1), S3.

Kosten, C. (1960) *Acustica* **10**, 245–250.

Koyasu, M. and Sato, K. (1957) *J. Acoust. Soc. Jpn.* **13**, 231–241 (in Japanese).

Kuhn, G. F. (1977) *J. Acoust. Soc. Am.* **80**, 661–664

Kuttruff, H. (1958) *Acustica* **8**, 273–280.

Kuttruff, H. (1991) *Room Acoustics*, 3rd edn. Elsevier Science Publishers, New York.

Lev. A. and Sohmer, H. (1972) *Arch. Klin. Exp. Ohr., Nas, u. Kehlk. Heilk.* **201**, 79–90.

Levin, M. J. (1964) *IEEE Trans. Information Theory*, **IT-10**, 95–97.

Ljung, L. and Soederstroem, T. (1983) *Theory and Practice of Recursive Identification*. MIT Press, Cambridge, MA.

London, A. (1945) *J. Res. Natl. Bur. Stad.* **42**, 605–615.

Lubman, D. (1969) *J. Acoust. Soc. Am.* **46**, 532–534.

Lubman, D. (1974) *J. Acoust. Soc. Am.* **56**, 1302–1304.

Lyon, R. H. (1969) *J. Acoust. Soc. Am.* **45**, 545–565.

Lyon, R. H. (1975) *Statistical Energy Analysis of Dynamical Systems: Theory and Applications*. MIT Press, Cambridge, MA.

Lyon, R. H. (1983) *J. Acoust. Soc. Am.* **73**, 1223–1228.

Lyon, R. H. (1984) *J. Acoust. Soc. Am.* **76**, 1433–1437.

Lyon, R. H. (1987) *Machinery Noise and Diagnostics*. Butterworths, Boston MA.

Lyon, R. H. (1991) *J. Acoust. Soc. Am.* **87** (suppl. 1) MM1 (S91).

Makino, S. and Kaneda, Y. (1992) *Trans. IEICE* **E75-A**, 1500–1508.

Makino, S., Kaneda, Y. and Koizumi, N. (1993) *IEEE Trans. Speech & Audio Processing* **1**, 101–108

Makita, Y. (1962) *EBU Review* **37A**, 102–108.

Mandelbrot, B. (1983) *The Fractal Geometry of Nature*. W. H. Freeman, New York.

Mehta, M. L. (1991) *Random Matrices*. Academic Press, San Diego.

Miyata, H. and Houtgast, T. (1991) *Proc. Eurospeech '91*, pp. 289–292.

Miyata, H., Nomura, H. and Houtgast, T. (1991) *Acustica* **73**, 200–207.

Miyoshi, M. and Kaneda, Y. (1988) *Trans. IEEE*, **ASSP36**, 145–152.

Miyoshi, M. and Kaneda, Y. (1991) *Noise Control Eng. J.* **36**, 85–90.

Miyoshi, M. and Koizumi, N. (1991) *Proc. Int. Symp. Active Control of Sound and Vibration*, Tokyo, pp. 217–222.

Miyoshi, M. and Koizumi, N. (1992) *Appl. Acoust.* **36**, 307–326.

Mori, T., Ando, Y. and Maekawa, J. (1987) *Report of Meeting of Acoust. Soc. Jpn.* 291–292 (in Japanese).

Morimoto, M., Shunto, M. and Maekawa, Z. (1983) *J. Architecture, Environ. Eng. (Trans. AIJ)* **4**, 64–69.

Morse, P. and Bolt, R. (1944) *Rev. Mod. Phys.* **16**, 69–150.

Morse, P. and Ingard, K. (1968) *Theoretical Acoustics*. Princeton University Press, Princeton, NJ.

Nagamatsu, H., Kasai, H. and Ando, Y. (1989) *Report of Meeting of Acoust. Soc. Jpn* 391–392.

Nagumo, J. and Noda, A. (1967) *Trans. IEEE*, **AC-12**, 282–287.

Nakabayashi, K. (1977) *J. Acoust. Soc. Jpn* **33**, 116–127 (in Japanese).

Nakajima, T. (1992) Speech intelligibility and clarity related to spatial-binaural factor for sound field in a room. PhD Dissertation, Kobe University.

Nakajima, T. and Ando, Y. (1991) *J. Acoust. Soc. Am.* **90**, 3173–3179.

Nakajima, T., Ando, Y. and Fujita, K. (1992) *J. Acoust. Soc. Am.* **92**, 1443–1451.

Nakayama, I. (1984) *Acustica* **54**, 217–221.

Nelson, P. A. and Elliott, S. J. (1991) *Active Noise Control*. Academic Press, London.

Nelson, P., Curtis, A., Elliott, S. and Bullmore, J. (1987) *J. Sound Vib.* **117**, 1–13.

Nelson, P., Hamada, H. and Elliott, S. (1992) *IEEE Trans.* **ASSP40**, 1621–1632.

Noiseux, D. U. (1970) *J. Acoust. Soc. Am.* **47**, 238–247.

Nomura, H., Miyata, H. and Houtgast, T. (1989) *Acustica* **69**, 151–155.

Nomura, H., Tohyama, M. and Houtgast, T. (1991) *J. Audio Eng. Soc.* **39**, 338–343.

Nomura, H., Miyata, H. and Houtgast, T. (1992) *J. Audio Eng. Soc.* **41**, 771–781.

Nomura, H., Miyata, H. and Houtgast, T. (1993) *Acustica* **77**, 253–261.

Nyquist, H. (1932) *Bell Syst. Tech. J.* **11**, 126–147.

Ochi, M. (1990) *Applied Probability and Stochastic Processes*. Wiley, New York.

Oppenheim, A. V. and Shafer, R. W. (1975) *Digital Signal Processing*. Prentice-Hall, Englewood Cliffs, NJ.

Oppenheim, A. V., Schafer, R. W. and Stockham Jr, T. G. (1968) *Proc. IEEE* **56**, 1264–1291.

Page, C. H. (1952) *J. Appl. Phys.* **23**, 103–106.

Papoulis, A. (1962) *The Fourier Integral and Its Applications*. McGraw-Hill, New York.

Pavic, G. (1976) *J. Sound Vib.* **49**, 221–230.

Pavic, G. (1981) *Proc. CETIM*, pp. 209–215. Prentice-Hall, Englewood Cliffs, NJ.

Pioneer Corporation (1992) *J. Acoust. Soc. Jpn* **48**, 275–278 (in Japanese).

Rabiner, L. R. and Shafer, R. W. (1978) *Digital Processing of Speech Signals*. Prentice-Hall, Englewood Cliffs, NJ.

Rice, S. O. (1945) *Bell Syst. Tech. J.* **24**, 46–108.

Richardson, M. H. and Formenti, D. L. (1985), *IMAC Proceedings*, pp. 28–31

Rihaczek, A. W. (1968) *IEEE Trans. Information Theory*, **IT-14**, 369–374.

Roderman, R. (1987) *IMAC Proceedings*, pp. 37–41.

Sabine, W. (1922) *Collected Papers on Acoustics*. Harvard University Press.

Sakamoto, N., Gotoh, T., and Kogure, T. et al. (1981) *J. Audio Eng. Soc.* **29**, 794–799.

Schroeder, M. R. (1954a) *Acustica* **4**, 456–468.

Schroeder, M. R. (1954b) *Acustica* **4** (suppl. 2)), 594–600.

Schroeder, M. R. (1962) *J. Acoust. Soc. Am.* **34**, 1819–1823.

Schroeder, M. R. (1965a) *Proceedings of 5th International Congress of Acoustics*, p. G31.

Schroeder, M. R. (1965b) *J. Acoust. Soc. Am.* **37**, 409–412.

Schroeder, M. R. (1964) *J. Acoust. Soc. Am.* **36**, 1718–1724.

Schroeder, M. R. (1970) *J. Acoust. Soc. Am.* **47**, 424–431.

Schroeder, M. R. (1979) *J. Acoust. Soc. Am.* **65**, 958–963.

Schroeder, M. R. (1981) *Acustica* **49**, 179–182.

Schroeder, M. R. (1985) *Number Theory in Science and Communication* (2nd enlarged edn). Springer-Verlag, Berlin.

Schroeder, M. R. (1989) *J. Audio Eng. Soc.* **37**, 795–808.

Schroeder, M. R. (1991) *Fractal, Chaos, Power Laws*. W. H. Freeman, New York.

Schroeder, M. R. and Atal, B. (1963) *IEEE Int. Conv. Rec.* **7**, 150–155.

Schroeder, M. R. and Hackman, D. (1980) *Acustica* **45**, 269–273.

Schroeder, M. R., Gottlob, D. and Siebrasse, K. (1974) *J. Acoust. Soc. Am.* **56**, 1195–1201.

Seraphim, H. P. (1961) *Acustica* **11**, 80–91.

Shannon, C. E. (1949) *Communication in the Presence of Noise*, Proc. IRE.

Shaw, E. A. G. (1966) *J. Acoust. Soc. Am.* **39**, 465–470.

Shaw, E. A. G. (1974) *J. Acoust. Soc. Am.* **56**, 1848–1861.

Shimada, S. and Hayashi, S. (1992) *Proc. FASE'92* (Fed. Acoust. Soc. Europe), pp. 157–160.

Singh, P. K., Ando, Y. and Kurihara, Y. (1994) *Acustica* **80**, 471–477

Skudrzyk, E. (1971) *The Foundations of Acoustics*. Springer-Verlag, Berlin.

Soong, T. T. (1973) *Random Differential Equations in Science*. Academic Press, London.

Sperry, R. W. (1974) *The Neurosciences: Third Study Program*. MIT Press, Cambridge, MA.

Stevens, S. (1957) *J. Acoust. Soc. Am.* **29**, 603–606.

Sugiyama, K. and Tohyama, M. (1989) *J. Acoust. Soc. Jpn* **45**, 527–533.

Suzuki, A. and Tohyama, M. (1988) *Rev. Electrical Communications Laboratories, NTT*, **37**, 155–158.

Suzuki, H. (1980) *J. Audio. Eng. Soc.* **28**, 570–574.

Suzuki, H. (1981a) *J. Acoust. Soc. Am.* **69**, 41–49.

Suzuki, H. (1981b) *J. Acoust. Soc. Am.* **70**, 1480–1487.

Suzuki, H., Oguro, S., Anzai, M. et al. (1990a) *J. Acoust. Soc. Jpn* (E) **11**, 343–349.

Suzuki, H., Oguro, S., Anzai, M. and Ono, T. (1991) *Proc. Inter-noise '91*, pp. 1037–1040.

Terada, T., Nakajima, H., Tohyama, M. and Hirata, Y. (1994) *Proc. IEEE-SP Int. Symp. TF/TS Analysis (Philadelphia)*, 429–432

Thomas, L. B. (1969) *Statistical Communication Theory*. Wiley, New York.

Thompson, A. M. and Thompson, G. C. (1988) *J. Neurosci. Meth.* **25**, 13–17.

Thompson, G. C. and Thompson, A. M. (1988) *J. Acoust. Soc. Am.* **84** (suppl. 1), S12.

Thurstone, L. L. (1927) Psychol. Rev. **34**, 273–289.

Tohyama, M. (1986) J. Sound Vib. **108**, 339–343.

Tohyama, M. (1993) Proc. Noise and Man. 93, Nice.

Tohyama, M. and Itow, T. (1974) Acustica **30**, 1–11.

Tohyama, M. and Lyon, R. H. (1989a) J. Acoust. Soc. Am. **86**, 1854–1863.

Tohyama, M. and Lyon, R. H. (1989b) J. Acoust. Soc. Am. **86**, 2025–2029.

Tohyama, M. and Suzuki, A. (1986a) J. Sound Vib. **111**, 391–398.

Tohyama, M. and Suzuki, A. (1986b). J. Acoust. Soc. Am. **80**, 828–832.

Tohyama, M. and Suzuki, A. (1987) J. Sound Vib. **119**, 562–564.

Tohyama, M. and Suzuki, A. (1989) J. Acoust. Soc. Am. **85**, 780–786.

Tohyama, M. and Yoshikawa, S. (1981) J. Acoust. Soc. Am. **70**, 1674–1678.

Tohyama, M., Suzuki, A. and Yoshikawa, S. (1977) Acustica **39**, 51–53.

Tohyama, M., Suzuki, A. and Yoshikawa, S. (1979) Acustica **42**, 184–186.

Tohyama, M., Imai, A. and Tachibana, H. (1989) J. Sound Vib. **128**, 57–69.

Tohyama, M., Suzuki, A. and Sugiyama, K. (1989) Proc. Int. Conf. Acoustics, Speech and Signal Processing (Glasgow), pp. 2037–2040.

Tohyama, M., Suzuki, A. and Sugiyama, K. (1991) IEEE Trans., **SP39** 246–248.

Tohyama, M., Lyon, R. H. and Koike, T. (1991) J. Acoust. Soc. Am. **89**, 1701–1707.

Tohyama, M., Lyon, R. H. and Koike, T. (1992a) J. Acoust. Soc. Am. **91**, 2805–2812.

Tohyama, M., Lyon, R. H. and Koike, T. (1992b) Inverse Problems in Engineering Mechanics, ed. M. Tanaka and H. D. Bui. Springer-Verlag, Berlin.

Tohyama, M., Lyon, R. H. and Koike, T. (1992c) 2nd. Int. Congr. Recent Developments in Air- and Structure-borne Sound and Vibration (Auburn), pp. 869–873.

Tohyama, M., Lyon, R. H. and Koike, T. (1993) Proc. Conf. Speech and Signal Processing (Minneapolis), pp. 157–160.

Tohyama, M., Lyon, R. H. and Koike, T. (1994) J. Acoust. Soc. Am. **95**, 286–296.

Ushiyama, S., Hirai, T., Tohyama, M. and Shimizu, Y. (1994) Technical Report of IEICE (Japan), EA94-4 (in Japanese)

Wang, H. and Itakura, F. (1989) IEEE ASSP Workshop on Applications of Signal Processing to Audio and Acoustics, Session 3-7.

Wang, H. and Itakura, F. (1992) IEICE Jpn Trans. **E75A**, 1474–1483.

Waterhouse, R. V. (1955) J. Acoust. Soc. Am. **27**, 247–258.

Waterhouse, R. V. (1958) J. Acoust. Soc. Am. **30**, 4–13.

Waterhouse, R. V. (1965) J. Acoust. Soc. Am. **37**, 921–923.

Waterhouse, R. V. (1963) J. Acoust. Soc. Am. **35**, 1144–1151.

Waterhouse, R. V. (1968) J. Acoust. Soc. Am. **43**, 1436–1444.

Weaver, H. J. (1989) Theory of Discrete and Continuous Fourier Analysis. Wiley, New York.

Wente, E. C. (1935) J. Acoust. Soc. Am. **7**, 123–126.

Wigner, E. (1965) Statistical Theory of Spectra: Fluctuations, ed. C. Porter, pp. 188–195. Academic Press, London.

Wigner, E. (1932) Physical Review **40**, 749–759.

Yamaha Corporation (1992) J. Acoust. Soc. Jpn **48**, 279–282 (in Japanese).

Yamasaki, Y. and Itow, T. (1989) J. Acoust. Soc. Jpn (E) **10**, 101–110.

Yamasaki, Y., Moriwake, K., Uno, T. et al. (1982) Proc. Autumn Meeting Acoust. Soc. Jpn, pp. 315–316.

Yanagawa, H. (1988) J. Acoust. Soc. Jpn **44**, 575–580.

Yanagawa, H. (1989) J. Acoust. Soc. Am. **86** (suppl. 1), S34.

Yanagawa, H., Yamasaki, Y. and Itow, T. (1988) J. Acoust. Soc. Am. **84** 1728–1733.

Yanagawa, H., Anazawa, T. and Itow, T. (1990) Acustica **71**, 230–232.

Yanagawa, H., Sato, T. and Matsushita, F. (1991) IEICE Tech. Rep. **EA91-53** (in Japanese).

Yegnanarayana, B. and Ramakrishna, B. S. (1975) J. Acoust. Soc. Am. **58**, 853–857.

Zwicker, E., Flottorp, G. and Stevens, S. S. (1957) J. Acoust. Soc. Am. **29**, 548–557.

INDEX

321